PROTECTION TECHNIQUES in ELECTRICAL ENERGY SYSTEMS

HELMUT UNGRAD
ABB Relays, Ltd.
Baden, Switzerland

WILIBALD WINKLER
Power System Automation and Protection
Silesian Technical University of Gliwice
Gliwice, Poland

ANDRZEJ WISZNIEWSKI
Power System Automation and Protection
Wrocław Technical University
Wrocław, Poland

Translated by
Peter G. Harrison

CRC Press
Taylor & Francis Group
Boca Raton London New York

CRC Press is an imprint of the
Taylor & Francis Group, an **informa** business

CRC Press
Taylor & Francis Group
6000 Broken Sound Parkway NW, Suite 300
Boca Raton, FL 33487-2742

First issued in paperback 2019

Originally published in German as *Schutztechnik in Elektroenergiesystemen*
©1992 Springer-Verlag, Inc.

ISBN-13: 978-0-8247-9660-0 (hbk)
ISBN-13: 978-0-367-40164-1 (pbk)

Library of Congress Cataloging-in-Publication Data

Ungrad, H. (Helmut)
    [Schutztechnik in Elektroenergiesystemen. English]
    Protection techniques in electrical energy systems / Helmut
Ungrad, Wilibald Winkler, Andrzej Wiszniewski ; translated by Peter
G. Harrison.
        p. cm.
    Translation of: Schutztechnik in Elektroenergiesystemen.
    Includes bibliographical references and index.
    ISBN 0-8247-9660-8 (hardcover : alk. paper)
    1. Electric power systems—Protection. I. Winkler, W.
(Wilibald) II. Wiszniewski, A. (Andrzej) III. Title.
TK1005.U48 1995
623.31'7—dc20   95-32278
                                                                      CIP

Visit the Taylor & Francis Web site at
http://www.taylorandfrancis.com

and the CRC Press Web site at
http://www.crcpress.com

Preface

In spite of careful design and maintenance, power systems and items of power system plant will always be subject to failure due to electrical or mechanical breakdown. In order to confine the consequences of such failures to the minimum possible, the defect has to be detected quickly, localized and isolated, and respective measures taken to ensure the continued supply of electrical energy to the consumers. These functions are performed automatically today by special protective devices and systems which constitute an important part of the auxiliary equipment in power plants and are the tools of the protection engineer.

This book presents the theoretical principles and current state of the art of protection engineering and also gives an outlook toward future developments. Correspondingly, the subject matter is dealt with in three parts:

- Part A explains the basic principles of protection engineering.

- Part B presents conventional analogue protection techniques as they are still being applied throughout the world today.

- Part C is devoted to the fundamentals of digital protection techniques which are already being introduced and will play an increasingly more important role in the future.

The structure of the book and the choice of the subject matter reflect the authors' many years of experience in development, research and training in the field of protection engineering. The material is presented in a way which makes the book equally suitable for use by students of electrical engineering and for the further training of electrical engineers. It is with this hope and intention in mind that the authors have written this book.

Helmut Ungrad · Wilibald Winkler · Andrzej Wiszniewski

Contents

CONTENTS

Part C: Digital Protection

Part A
Fundamentals of Protection
Engineering

1

Introduction

1.1   The purpose of protection engineering

The purpose of protection equipment is to minimize the effects of faults on electrical power systems, which unfortunately can never be entirely avoided. In this context, an electrical power system is considered as all the plant required to generate, transmit and distribute electrical power, i.e. generators, power transformers, lines and cables, circuit-breakers, instrument transformers etc. The faults which occur can be the result of external or internal influences (e.g. lightning strikes, resulting in an overload). Protection engineering is thus an extremely important part of the secondary electrical plant and of decisive significance for the reliable operation of electrical power systems.

Since the damage a fault can cause is mainly dependent on its duration, it is necessary for the protection devices to operate as quickly as possible. They must, however, operate absolutely selectively in order to isolate only the faulty item of plant. They must also operate reliably, i.e. there should be a tendency neither to overfunction nor underfunction. These requirements are in part contradictory and finding the right compromise is one of the main tasks of the protection engineer (see Section 1.2).

The use of automatic equipment to minimize the consequences of a fault by switching in reserve plant after the defective item of plant has been tripped is becoming more widespread.

Among the most important consequences of a fault are:
- damage to plant due to the dynamic effects of the fault current
- damage to plant due to the thermal effects of the current
- loss of system stability
- loss of supply to loads, also during downtime for repairs
- danger to life

1.2 Requirements to be fulfilled by protection devices

The following basic demands must always be fulfilled:

1. **Selectivity:** The ability to isolate only the defective plant from the rest of the system; this can be achieved by the following methods:
   - time grading, i.e. the protection device nearest the fault trips the fastest and all the others between it and the power source successively slower, delayed by either a fixed "definite" time or a time inversely proportional to the level of fault current. Application: overcurrent and distance protection.
   - amplitude and/or phase comparison of the currents on both sides of the protected unit. Application: pilot wire and differential protection.
   - determination of direction (of fault power flow) on both sides of the protected unit by comparing derived signals. Application: directional comparison protection and distance protection with a communications channel.

   The last two methods are also referred to as "unit protection", because they only detect faults between the c.t's on both sides of the protected unit and do not provide back-up protection for neighbouring plant.

2. **Reliability:** The ability of a protection device to fulfil its purpose throughout its operational life [1.1] ... [1.5]. A distinction is made between:
   - **dependability:** The assurance that the protection device will perform its designated function and selectively trip the protected item of primary plant in the event of a fault.
   - **security:** The assurance that the protection device will not trip unless there is a fault on the protected item of primary plant.

   The following term is also used:
   - **availability:** The ratio of the time that a protection device is actually serviceable to the total time it is in operation.

High availability of protection equipment can be achieved by the following:
- high technical quality of all the components in the protection chain (Fig. 1.1a), i.e. instrument transformers, protection equipment, auxiliary supply batteries and distribution system, any communication channels, circuit-breaker and its mechanism and all the interconnections between these components.

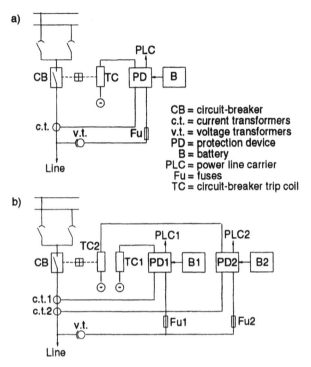

CB = circuit-breaker
c.t. = current transformers
v.t. = voltage transformers
PD = protection device
B = battery
PLC = power line carrier
Fu = fuses
TC = circuit-breaker trip coil

Fig. 1.1:   A simple (a) and a redundant (b) protection scheme for an HV line

- optimum design of the protection scheme, e.g. installation of a redundant protection chain (duplication) for the protection of important HV lines according to Fig. 1.1b.
- continuous self-monitoring of the protection devices.
- carefully carried out acceptance and commissioning tests, periodic testing, automatic testing routines etc.

3. **Operating speed:** The time between the incidence of a fault and the trip command being issued to the circuit-breaker by the protection is determined by the power system configuration and in the case of modern protection devices is typically one period of the power system frequency or fractions of a period [1.6]. Amongst other things, fault clearance times which are too long increase the thermal stress on the protected plant and endanger power system stability. The required speed of the protection is also dependent on the significance of the item of plant for the power system and its insured value. The total fault clearance time is the sum of protection tripping time and the rupture time (including arc extinguishing time) of the circuit-breaker. Since today's circuit-breakers take two and more periods of the power system frequency to

interrupt the current, fault clearance times in the order of about 50 ... 80 ms are possible and these are the times used for HV power systems or parts of power systems close to large power plants. Times of 100 ms and higher are typical in MV systems. Preventive protection devices, which do not trip automatically themselves but give warning when dangerous situations arise, may operate with correspondingly longer time delays.

4. **Sensitivity:** This is the ability of a protection device to react even to relatively small deviations of the monitored (normally electrical) variable from its normal load value due to a disturbance on the primary system.

1.3 Basic structure of protection systems

Figure 1.2 shows the flow of information between the protected unit and the individual links of the protection chain. The data describing the physical variables being monitored are conveyed from the primary system to the protection device PD and appropriately adjusted by the transducers IT, i.e. c.t's, v.t's, temperature feelers etc. The protection device can take the form of a single protection relay performing only a single measurement or - in the case of an especially valuable item of plant such as a large generator - it can be a protection system comprising several separate functions.

A very important link in the protection chain from the point of view of reliability is the auxiliary supply AS. This is generally the station battery which also supplies the circuit-breaker tripping coils CC and the display and recording devices D&R. The auxiliary supply units usually installed in the protection devices to derive the internal supply voltages also belong in this category. Continuously

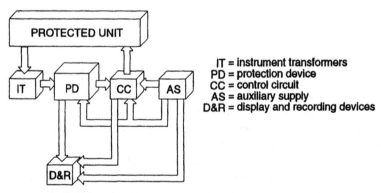

IT = instrument transformers
PD = protection device
CC = control circuit
AS = auxiliary supply
D&R = display and recording devices

Fig. 1.2:   Basic structure of a protection system

monitoring all the auxiliary supply voltages is urgently recommended and monitors to do this are installed in modern protection devices.

A protection device that has detected a fault on the item of primary plant it is protecting issues a trip command to the circuit-breaker or breakers, which then isolate it from the rest of the system.

Where the possibility exists of the fault correcting itself (e.g. a fault on an overhead line ignited by a lighting strike) after it has been isolated for a short time (a few 100 ms), an attempt can be made following a delay to automatically reconnect the item of plant tripped to restore the normal status of the power system.

In the case of digital systems which combine protection and control functions, what is referred to as a "feeder module" can perform metering, interlocking and controller functions in addition to the protection functions. The latter operate nevertheless completely independently, i.e. they continue to function should the station control system or communication link fail. The feeder modules are connected to a station control system by a (serial) data link [1.7].

Adaptive protection systems vary their settings or configurations to suit changing power system operating conditions [1.8].

2

The Principal Faults on Electrical Power Systems and System Components

2.1 Phase faults

Because of the high dynamic and thermal stresses on plant and materials caused by fault currents, phase faults are the most dangerous kinds of faults which can occur on an electrical power system. By way of example, Table 2.1 lists rated currents and phase fault currents for three-phase faults directly at the terminals of synchronous generators. Phase fault currents of ten times rated current are not infrequent. Similar conditions prevail at nodes of power concentration on the power system.

A distinction is made between the following main kinds of faults (Fig. 2.1):

    (a) three-phase faults
    (b) phase-to-phase faults
    (c) phase-to-phase-to-ground faults
    (d) ground faults
    (e) cross-country faults

Table 2.1 Three-phase fault currents in relation to the rated currents of synchronous generators

P_N [MW]	U_N [kV]	I_N [kA]	S''_{F3ph} [MVA]	I''_{F3ph} [kA]	I''_{F3ph}/I_N [--]
30	11.8	1.85	300	14.7	7.9
120	10...15	5.4...9.2	1000	57.8...38.5	10.7
300	12.5...22.5	9...18.5	2500	115.6...64.2	12.8
600	20...25	16...21.5	4000	115.6...92.4	7.2
1200	22...27.5	30...40	7000	184...147	6.1

The initial values of AC fault currents can be calculated, for example, according to VDE 0102 [2.1]. The corresponding formulas for the faults of (a) to (d) in Fig. 2.1 are summarized in Table 2.2.

It is important to know the fault level at the relay location in order to determine the behaviour of the protection at the steady-state fault level; this can be calculated according to the principle of symmetrical components [2.2] ... [2.6].

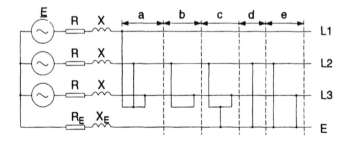

Fig. 2.1: The different kinds of phase and ground faults
a) three-phase   b) phase-to-phase   c) phase-to-phase-to-ground
d) ground fault   e) cross-country fault

Table 2.2   Formulas for calculating initial AC fault currents in three-phase power systems with rated voltages of 1 kV and higher [2.1]

Kind of fault	Fault current formula
Three-phase fault	$I''_{F3ph} = \dfrac{1.1U_N}{\sqrt{3}\lvert \underline{Z}_1 \rvert}$
Phase-to-phase fault	$I''_{F2ph} = \dfrac{1.1U_N}{\lvert \underline{Z}_1 + \underline{Z}_2 \rvert}$
Phase-to-phase-to-ground	$I''_{F2phE} = \dfrac{\sqrt{3}(1.1U_N)}{\left\lvert \underline{Z}_1 + \underline{Z}_0 + \underline{Z}_0 \left(\dfrac{\underline{Z}_1}{\underline{Z}_2} \right) \right\rvert}$
Ground fault	$I''_{E/F} = \dfrac{\sqrt{3}(1.1U_N)}{\lvert \underline{Z}_1 + \underline{Z}_2 + \underline{Z}_0 \rvert}$

Assuming the equivalent circuits given in Fig. 2.2, the positive, negative and zero-sequence components of the fault current at the relay location A and the fault location F can be represented as follows (the conductor L1 is taken as reference):

$$I_{1A} = \frac{\underline{Z}_{1QB}}{\underline{Z}_{1QA} + \underline{Z}_{1AF} + \underline{Z}_{1QB}} I_{1F} + I_L$$

$$I_{2A} = \frac{\underline{Z}_{2QB}}{\underline{Z}_{2QA} + \underline{Z}_{2AF} + \underline{Z}_{2QB}} I_{2F} \qquad (2.1)$$

$$I_{0A} = \frac{\underline{Z}_{0QB}}{\underline{Z}_{0QA} + \underline{Z}_{0AF} + \underline{Z}_{0QB}} I_{0F}$$

The indices A and F denote the relationship to the relay location (A) and the fault location (F), 1, 2 and 0 the positive, negative and zero-sequence components, Q the source impedance and I_L the load current at the relay location prior to the fault. All the other variables can be seen from Fig. 2.2.

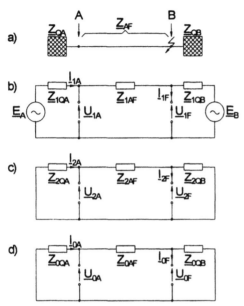

Fig. 2.2: Network diagram a) and equivalent circuits for positive, negative and zero-sequence systems b), c) and d) where A = relay location, F = fault location and Z_{1N}, Z_{2N}, Z_{0N} = positive, negative and zero-sequence systems

The symmetrical components for the voltage at the relay location are given by

$$U_{1A} = U_L - \frac{Z_{1QB}}{Z_{1QA} + Z_{1AF} + Z_{1QB}} I_{1F}$$

$$U_{2A} = -\frac{Z_{2QB}}{Z_{2QA} + Z_{2AF} + Z_{2QB}} I_{2F} \qquad (2.2)$$

$$U_{0A} = -\frac{Z_{0QB}}{Z_{0QA} + Z_{0AF} + Z_{0QB}} I_{0F}$$

U_L is the load voltage at the relay location prior to the fault.

Table 2.3 lists the positive, negative and zero-sequence components of the fault current at the fault location for the most important kinds of faults. The equations are based on the equivalent circuits for symmetrical components [2.4], from which the phase currents and voltages at the relay location can be derived by applying the following transformations:

$$\underline{I}_{L1A} = \underline{I}_{1A} + \underline{I}_{2A} + \underline{I}_{0A}$$

$$\underline{I}_{L2A} = a^2\underline{I}_{1A} + a\underline{I}_{2A} + \underline{I}_{0A} \tag{2.3}$$

$$\underline{I}_{L2A} = a\underline{I}_{1A} + a^2\underline{I}_{2A} + \underline{I}_{0A}$$

$$\underline{U}_{L1A} = \underline{U}_{1A} + \underline{U}_{2A} + \underline{U}_{0A}$$

$$\underline{U}_{L2A} = a^2\underline{U}_{1A} + a\underline{U}_{2A} + \underline{U}_{0A} \tag{2.4}$$

$$\underline{U}_{L2A} = a\underline{U}_{1A} + a^2\underline{U}_{2A} + \underline{U}_{0A}$$

where $a = e^{j120°}$ and $a^2 = e^{j240°}$

The transient excursion of the fault current for a three-phase fault without fault resistance can be determined from the equivalent circuit of Fig. 2.3.

Table 2.3: Symmetrical components of the current at the fault location for different kinds of faults

Kind of fault	Pos.-sequence current I_{1F}	Neg.-sequence current I_{2F}	Zero-sequence current I_{0F}
Three-phase fault	$\dfrac{\underline{E}}{\underline{Z}_1}$	0	0
Phase-to-phase fault L2-L3	$\dfrac{\underline{E}}{\underline{Z}_1 + \underline{Z}_2}$	$-\dfrac{\underline{E}}{\underline{Z}_1 + \underline{Z}_2}$	0
Phase-to-phase-to-ground fault L2-L3-E	$\dfrac{(\underline{Z}_2 + \underline{Z}_0)\underline{E}}{\underline{Z}_1\underline{Z}_2 + \underline{Z}_2\underline{Z}_0 + \underline{Z}_0\underline{Z}_1}$	$\dfrac{-\underline{Z}_0\underline{E}}{\underline{Z}_1\underline{Z}_2 + \underline{Z}_2\underline{Z}_0 + \underline{Z}_0\underline{Z}_1}$	$\dfrac{-\underline{Z}_2\underline{E}}{\underline{Z}_1\underline{Z}_2 + \underline{Z}_2\underline{Z}_0 + \underline{Z}_0\underline{Z}_1}$
Ground fault L1-E	$\dfrac{\underline{E}}{\underline{Z}_1 + \underline{Z}_2 + \underline{Z}_0}$	$\dfrac{\underline{E}}{\underline{Z}_1 + \underline{Z}_2 + \underline{Z}_0}$	$\dfrac{\underline{E}}{\underline{Z}_1 + \underline{Z}_2 + \underline{Z}_0}$
$\underline{Z}_1 = \dfrac{(\underline{Z}_{1QA} + \underline{Z}_{1AF})\underline{Z}_{1QB}}{\underline{Z}_{1QA} + \underline{Z}_{1AF} + \underline{Z}_{1QB}}$;	$\underline{Z}_2 = \dfrac{(\underline{Z}_{2QA} + \underline{Z}_{2AF})\underline{Z}_{2QB}}{\underline{Z}_{2QA} + \underline{Z}_{2AF} + \underline{Z}_{2QB}}$;	$\underline{Z}_0 = \dfrac{(\underline{Z}_{0QA} + \underline{Z}_{0AF})\underline{Z}_{0QB}}{\underline{Z}_{0QA} + \underline{Z}_{0AF} + \underline{Z}_{0QB}}$	

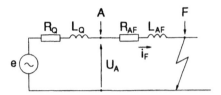

Fig. 2.3:   Equivalent circuit of a solid three-phase fault

Assuming the behaviour of the source voltage in relation to time is given by the following equation

$$e = \hat{E} \cos(\omega t + \beta) \tag{2.5}$$

the fault current becomes

$$i_F = \hat{I}\left[e^{-\frac{t}{T_N}} \cos\alpha - \cos(\omega t + \alpha)\right] \tag{2.6}$$

$$\underbrace{\qquad\qquad}_{\text{DC comp.}} \underbrace{\qquad\qquad}_{\text{AC comp.}}$$

where

$$\hat{I} = \frac{\hat{E}}{\sqrt{R^2 + (\omega L)^2}} \quad \text{in which} \quad R = R_Q + R_{AF} \quad \text{and} \quad L = L_Q + L_{AF}$$

T_N = L/R of the power system
α = β - arc $\tan(\omega L/R)$
β = phase-angle of the source voltage at fault incidence

It can be seen from equation (2.6) that the DC component which decays with a time constant of T_N becomes a maximum at $\alpha = 0$.
The following then applies

$$i_F = \hat{I}\left[e^{-\frac{t}{T_N}} - \cos\omega t\right] \tag{2.7}$$

At medium voltages, the power system time constant T_N of the DC component in the fault current is of the order of tens of milliseconds, but in HV and EHV systems and also close to large generators it can reach a few hundred milliseconds [2.7]. The time constant is reduced by any resistance at the fault location (arc resistance, earth wire resistance etc.), which is always the case for ground faults and phase-to-phase-to-ground faults.

The characteristic of the transient voltage at the fault location in relation to time is given by

$$u = \hat{U}\left[K_u\, e^{-\frac{t}{T_N}}\cos\alpha - \cos(\omega t + \alpha + \varphi)\right] \qquad (2.8)$$

$$\underbrace{\phantom{K_u\, e^{-\frac{t}{T_N}}\cos\alpha}}_{\text{DC comp.}} \quad \underbrace{}_{\text{AC comp.}}$$

where

$$u = \hat{U}\sqrt{\frac{R_{AF}^2 + (\omega L_{AF})^2}{R^2 + (\omega L)^2}} \qquad\qquad \varphi = \arctan\left(\frac{\omega L_{AF}}{R_{AF}}\right)$$

$$K_u = \frac{R_{AF}\left(1 - \dfrac{R L_{AF}}{L R_{AF}}\right)}{\sqrt{R_{AF}^2 + (\omega L_{AF})^2}} \qquad\qquad\qquad (2.9)$$

It follows that the initial value of the DC component is dependent on the factor K_u. The DC component becomes zero for $L_Q/R_Q = L_{AF}/R_{AF}$, but is generally of a low value in practice.

Apart from the DC component, both fault current and voltage at the relay location also contain proportions of HF generated by the capacitances of the line. For a greatly simplified equivalent circuit with a single lumped capacitance (Fig. 2.4) and assuming $L_Q/R_Q = L_{AF}/R_{AF}$, the transient characteristic of the voltage at the relay location for a source voltage according to equation (2.5) becomes

$$U_A = \hat{E}\left[\frac{L_{AF}}{L_Q + L_{AF}}\cos(\omega t + \beta) + \frac{L_Q}{L_Q + L_{AF}}e^{-\frac{t}{T_P}}\cos\beta\,\cos(\omega_p t)\right] \qquad (2.10)$$

$$\underbrace{\phantom{\frac{L_{AF}}{L_Q + L_{AF}}\cos(\omega t + \beta)}}_{\text{fundamental}} \quad \underbrace{\phantom{\frac{L_Q}{L_Q + L_{AF}}e^{-\frac{t}{T_P}}\cos\beta\,\cos(\omega_p t)}}_{\text{HF oscillation}}$$

where

$$T_p = \frac{2L_{AF}}{R_{AF}} \qquad\qquad \omega_p = \sqrt{\frac{L_Q + L_{AF}}{C\, L_Q\, L_{AF}}}$$

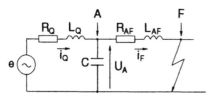

Fig. 2.4:   Simplified equivalent circuit including the ground capacitance C

Assuming the amplitudes of the partial currents i_Q and i_{AF} to be very low prior to the fault and the influence of the resistances on the amplitudes of the AC components to be negligible, the current at the relay location becomes

$$i_Q = \hat{i}\left[e^{-\frac{t}{T_N}}\cos\alpha - \cos(\omega t + \alpha) - h\, e^{-\frac{t}{T_P}}\sin\alpha\,\sin(\omega_p t)\right] \qquad (2.11)$$

$$\underbrace{\phantom{e^{-\frac{t}{T_N}}\cos\alpha}}_{\text{DC comp.}} \quad \underbrace{}_{\text{fundamental}} \quad \underbrace{\phantom{h\,e^{-\frac{t}{T_P}}\sin\alpha\,\sin(\omega_p t)}}_{\text{HF oscillation}}$$

where

$$\hat{i} = \frac{\hat{E}}{\omega(L_Q + L_{AF})} \qquad h = \frac{\omega}{\omega_p}$$

An analysis of equation (2.10) shows that in extreme cases the amplitude of the exponentially decaying HF component can equal the amplitude of the fundamental (for $\beta = 0$); the initial amplitude of the HF oscillation in the current i_Q on the other hand is always much smaller than the amplitude of the fundamental since $\omega < \omega_p$. It should also be noted that for $\alpha = 0$ the initial value of the DC component becomes a maximum, while the amplitude of the HF oscillation becomes zero. The situation is reversed for $\alpha = \pi/2$. The frequencies contained in the decaying oscillatory components depend on the parameters of the fault loop (source and line impedances); for HV and EHV lines these can be of the order of a few hundred Hz to a few tens of kHz.

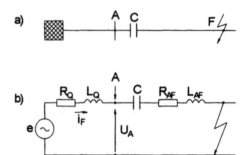

Fig. 2.5:   Principle of series compensation a) and a simplified equivalent circuit for a
            three-phase fault b)

The transient fault current and voltage at the relay location in the case of series compensated line (Fig. 2.5) contain a decaying oscillatory component instead
of a DC component. At this juncture, only the excursion of the fault current in
relation to time will be analysed; for the source voltage according to equation
(2.5), the procedure of [2.8] results in the expression

$$i_F = \hat{I}\left[\cos(\omega t + \alpha) - N\cos(\lambda t + \varphi)e^{-\frac{t}{T'_N}}\right] \qquad (2.12)$$

where

$$\varphi = \arctan\left(\frac{X_1}{R_1}\right); \qquad T'_N = \frac{2X_1}{\omega R_1}$$

$$N = \sqrt{\cos^2\alpha + \left(\frac{\lambda}{\omega}\right)^2 \sin^2\alpha}; \qquad \lambda \approx \omega\sqrt{\frac{X_C}{X_1}} \qquad (2.13)$$

and

$$X_1 = X_Q + X_C + X_{AF}; \qquad R_1 = R_Q + R_{AF}$$

It follows from equation (2.13) that $N = 1$ for $\alpha = 0$ and $N = \lambda/\omega$ for
$\alpha = 90°$.

Depending on the parameters of the transmission system (degree of compensation, fault location, source impedance etc.), the frequency of the decaying
oscillation can be in the range of 15 ... 150 Hz (for a rated frequency of 50 Hz).
The time constant T'_N on the other hand lies between 100 and 400 ms for a solid
three-phase fault. The values of frequency and time constant are lower for a
ground fault.

2.2 Ground faults in ungrounded or impedance grounded systems

Ground faults can also occur in ungrounded or high resistance or inductively grounded systems. The latter are systems which are grounded via a Petersen coil tuned to compensate the capacitive ground fault currents.

The current flowing in the event of a ground fault in a system of this kind is determined by the nature of the connection between the star point of the power system and ground and the other parameters of the power system, but is usually appreciably less than a ground fault in a solidly grounded system and in most cases less than the rated current of the plant.

The steady-state fault current I_{CE} at the fault location for a ground fault in an ungrounded system (Fig. 2.6a) is given by

$$\underline{I}_{CE} = \sqrt{3}\,j\omega\,C_E\,U_N \tag{2.14}$$

where

C_E = the ground capacitance of the metallically connected part of the power system

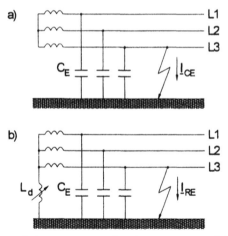

Fig. 2.6: An ungrounded power system a) and a system grounded via a Petersen coil b)

The transient fault current i_{CE} at the fault location can be determined with the aid of the equivalent circuit of Fig. 2.7 [2.9]. According to Thevenin

$$i_{CE} = i_a + i_b$$

The corresponding partial currents for a source voltage according to equation (2.5) are given by

$$i_a = -\hat{I}_a \left\{ \cos(\omega t + \alpha) + e^{-\frac{t}{T_a}} \left[\frac{\omega_a}{\omega} \sin\alpha \; \sin(\omega_a t) - \cos\alpha \; \cos(\omega_a t) \right] \right\} \qquad (2.15)$$

$$i_b = -\hat{I}_b \left\{ \cos(\omega t + \alpha) + e^{-\frac{t}{T_b}} \left[\frac{\omega_b}{\omega} \sin\alpha \; \sin(\omega_b t) - \cos\alpha \; \cos(\omega_b t) \right] \right\} \qquad (2.16)$$

where

$$\hat{I}_b = 0.5\hat{I}_a = \omega C_E E; \qquad\qquad T_a = \frac{3L_a + L_b}{3R_a + R_b}$$

$$\omega_a = \frac{1}{\sqrt{C_E(3L_a + L_b)}}; \qquad\qquad T_b = \frac{L_b}{R_b}$$

$$\omega_b = \frac{1}{\sqrt{C_E L_b}}; \qquad\qquad \alpha = \beta + \frac{\Pi}{2} \qquad\qquad (2.17)$$

From equations (2.15) and (2.16) it can be seen that two damped oscillations occur in the ground fault current. The component with the lower angular velocity ω_a is the physical consequence of charging the ground capacitances of the healthy conductors via the source inductance; the corresponding time constant is approximately 10^{-2} s and the frequency does not exceed 1 kHz. The second component with the higher angular velocity ω_b arises from discharging the ground capacitance of the faulted conductor; the time constant in this case is approximately 10^{-3} s and the frequency lies between a few kHz and a few tens of kHz. The amplitudes are several times that of the fundamental.

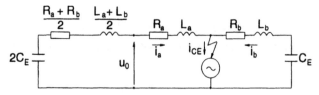

Fig. 2.7:   Equivalent circuit for determining the excursions of the transient currents and voltages in ungrounded power systems

The transient behaviour of the voltage u_0 can be determined from the equivalent circuit of Fig. 2.7. Based on the reasonable assumption that L_a is much greater than L_b,

$$u_0 = \hat{E}\left[\cos(\omega t + \beta) + \frac{2}{3}\sin\alpha\,\cos(\omega_a t)\,e^{-\frac{t}{\tau_a}} + \frac{2}{3}\left(\frac{\omega}{\omega_a}\right)\cos\alpha\,\sin(\omega_a t)\,e^{-\frac{t}{\tau_a}}\right] \quad (2.18)$$

It is clear that the voltage u_0 only contains a single damped oscillation with the angular velocity ω_a, the amplitude of which cannot exceed $2/3$ of the amplitude of the fundamental.

However, re-ignition often occurs in such power systems which causes the voltage to swing up to high levels. In fact, it can reach as much as 3.5 U_{ph} during "intermittent ground faults" of this kind, with the result that a flashover occurs at another location and the initial ground fault evolves into a cross-country fault.

The maximum voltage reached in an inductively grounded system is 2.5 U_{ph}.

In the case of a ground fault in a system with a Petersen coil (Fig. 2.6b), the stead-state fault current at the fault location depends on the tuning of the Petersen coil in relation to the ground capacitance of the power system. If the coil is ideally tuned, i.e. for $L_d = 1/3\omega^2 C_E$, a residual resistive current I_{RE} flows at the fault location as given by the equation

$$I_{RE} = 3\underline{E}\,R_\alpha\,\omega^2\,C_E^2 \quad (2.19)$$

where $R\alpha$ = resistance of the Petersen coil.

In practice

$$I_{RE} \approx (0.03 + 0.1)I_{CE} \quad (2.20)$$

Providing this current is lower than the natural arc extinguishing current, an arcing fault will extinguish on its own. Since the majority of faults in overhead line systems are arcing faults, this technique automatically eliminates them and is the reason why it is so widespread in MV systems.

In extensive MV systems or in HV systems, the above current is usually higher than the extinguishing limit of the arc (approx. 30 A) and therefore Petersen coils are not necessarily applicable.

The transient behaviour of the fault current immediately after the incidence of a ground fault in a system with a Petersen coil is similar to that in an ungrounded system. Only after some little time does the balancing process change as a consequence of the Petersen coil [2.10, 2.11].

2.3 Open-circuit lines

Open-circuit lines are caused either by an overhead line parting or the pole of a circuit-breaker not fully closing. The resulting load imbalance on generators and motors leads to an NPS component in the stator current, which rotates at twice the system frequency in the opposite direction in relation to the rotor and causes additional eddy current losses and therefore an additional temperature rise in the rotor. Amongst other things this temperature rise can produce welding of the rotor heads and synchronous motors can also fall out-of-step.

The magnitude of the NPS component in the stator current of a synchronous generator depends on the configuration of the power system. For a simple transmission system according to Fig. 2.8a, the value can be determined from the equivalent circuit of Fig. 2.8b [2.15]. For the interruption of phase L1 and a power transformer group of connection Yd 11

$$\underline{I}' = -\frac{I(\underline{Z}'_1 + \underline{Z}''_1)(\underline{Z}'_0 + \underline{Z}''_0)e^{-j30°}}{(\underline{Z}'_1 + \underline{Z}''_1)(\underline{Z}'_2 + \underline{Z}''_2 + \underline{Z}'_0 + \underline{Z}''_0) + (\underline{Z}'_2 + \underline{Z}''_2)(\underline{Z}'_0 + \underline{Z}''_0)} \tag{2.21}$$

where

$$\underline{I} = \frac{\underline{E}' - \underline{E}''}{\underline{Z}'_1 + \underline{Z}''_1} \tag{2.22}$$

$$\underline{Z}'_1 = \underline{Z}_{1G} + \underline{Z}_{1T} + \underline{Z}'_{1L}$$

$$\underline{Z}'_2 = \underline{Z}_{2G} + \underline{Z}_{2T} + \underline{Z}'_{2L} \tag{2.23}$$

$$\underline{Z}'_0 = \underline{Z}_{0T} + \underline{Z}'_{0L}$$

$$\underline{Z}''_1 = \underline{Z}''_{1L} + a\underline{Z}_{1S}$$

$$\underline{Z}''_2 = \underline{Z}''_{2L} + a\underline{Z}_{2S} \tag{2.24}$$

$$\underline{Z}''_0 = \underline{Z}''_{0L} + a\underline{Z}_{0S}$$

The parting of an overhead line is usually accompanied by a ground fault on the interrupted conductor. The corresponding equivalent circuit can be obtained and the NPS component determined by linking the symmetrical component systems [2.3], [2.6], [2.15], [2.16]. The following relationship results for the transmission system shown in Fig. 2.8a

$$\underline{I}'_2 = -\frac{\underline{E}'\,\underline{Z}'_0\,e^{-j30°}}{\underline{Z}'_1\underline{Z}'_2 + \underline{Z}'_1\underline{Z}'_0 + \underline{Z}'_2\underline{Z}'_0} \tag{2.25}$$

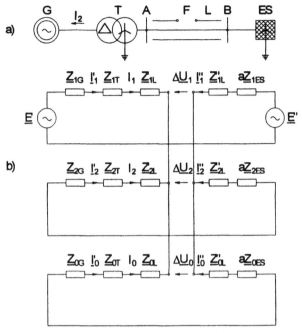

Fig. 2.8:   Transmission system a) and equivalent circuit b) for determining the NPS component in the generator stator current

Figure 2.9 shows the relative NPS content I_2 referred to the rated current of a 400 MW generator for the case of the two faults under consideration on a 100 km long 400 kV line [2.15]. It can be seen that as long as the conductor is simply interrupted and a ground fault does not occur as well, the value of the NPS component is uninfluenced by the fault location and is less than the rated current of the generator. Furthermore, the parameters of the transmission system have no significant influence on I_2. This is quite different when the broken conductor falls onto the ground and causes a ground fault.

A load imbalance also results from overhead lines which are not transposed, but this is usually within the level permissible for the generators.

The NPS load which can be tolerated by a turbo-alternator for a limited period is described by the equation

$$\left(\frac{I_2}{I_{NG}}\right)^2 t = K_G \tag{2.26}$$

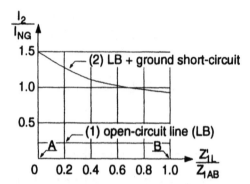

Fig. 2.9: NPS component in the stator current of a synchronous generator resulting from an interrupted conductor (1) and interrupted conductor and simultaneous ground fault (2)

Z'_{1L} = positive-sequence impedance between station A and fault location F

Z_{1AB} = positive-sequence impedance of the entire line (see Fig. 2.8a)

where K_G is determined by the design of the generator and the cooling system. The value of K_G lies between 2.5 and 60 s, the lower values applying to generators with higher ratings (600 ... 1500 MW) [2.17]. Figure 2.10 shows the permissible duration for an NPS load in relation to the level of the NPS component for three different turbo-alternators. As a general guide, the interruption of one

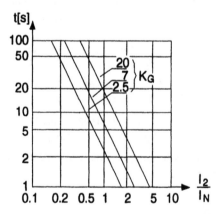

Fig. 2.10: Permissible duration of a load imbalance in relation to the relative value of the NPS component of the current for turbo-alternators

conductor at rated current produces an NPS current I_2/I_N of about 57 % in the other two phases.

2.4 Interturn faults in machines

An interturn fault is an insulation breakdown between turns of the same phase or between parallel windings belonging to the same phase. This kind of fault is particularly feared on synchronous generators, HV motors and power transformers. The cause is usually an atmospheric overvoltage or mechanical damage of the insulation.

Interturn faults on synchronous generators can befall both stator and rotor, but in the case of the stator they are only possible on generators of relatively low power with several rods in each groove (Fig. 2.11a), on generators with two stator windings per phase (Fig. 2.11b) or on HV motors. There is a high probability of an interturn fault evolving into a ground fault [2.3, 2.12]. The current in the shorted turn W_s can reach extremely high values and the winding can be completely destroyed. In the case of generators with two windings per phase, an interturn fault produces balancing currents of high current density, which are an extreme hazard for the winding.

Large generators generally only have a single winding rod per groove and therefore interturn faults are only possible in the winding head region.

Interturn faults on the rotor winding can produce the following symptoms [2.13]:

- high excitation current requirement which is automatically compensated by the voltage regulator

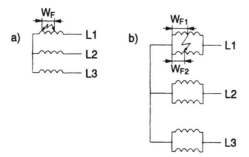

Fig. 2.11: Interturn faults on the stator windings of synchronous generators with one winding a) and two windings b) per phase

- machine runs less smoothly, because of the asymmetry of the excitation curve
- magnetization of the shaft due to the asymmetrical flux
- bearing damage due to currents flowing in the bearings, which in certain circumstances can result in a lowering of the shaft train and consequential blad damage in the prime-mover.

Interturn faults on power transformers can result from atmospheric overvoltages, overvoltages accompanying ground faults or deterioration of the insulation due to the chemical influence of the transformer oil. Since the shorted turn encompasses the entire flux, the high current flowing in it produces high dynamic forces which are frequently strong enough to destroy the winding. The interturn fault current depends on the number of turns shorted and can be several times higher than the rated current of the winding, while the phase current is only 2 ... 3 I_N [2.14].

2.5 Overload

Electrical plant can be thermally overloaded either by exceeding the maximum permissible load currents - usually over a prolonged period - and/or by a reduced cooling effectiveness. The temperature rise ϑ in an homogeneous body produced by a load Q can be derived by simplification from the familiar thermal equation [2.19]

$$Q \, dt = c_1 \, d\vartheta + c_2 \, \vartheta \, dt \tag{2.27}$$

gene-   stored   dissipated
rated

quantities of heat

The following solution is obtained

$$\vartheta = \vartheta_{max}\left(1 - e^{-\frac{t}{T_t}}\right) \tag{2.28}$$

where

ϑ_{max} $= Q/c_2$ = temperature limit

T_t $= c_1/c_2$ = thermal time constant of the plant

c_1, c_2 = coefficients which take account of the heat storage and dissipation capabilities of the plant

Designating the continuously permissible temperature of an item of plant to be ϑ_z and the continuously permissible load current I_z, the additional temperature rise resulting from the overload current I is $(I/I_z)^2$, i.e.

$$\vartheta = \vartheta_z \left(1 - e^{-\frac{t}{T_t}}\right)\left(\frac{I}{I_z}\right)^2 \tag{2.29}$$

The continuously permissible temperature ϑ_z is reached after the limit time t_g

$$t_g = T_t \ln \frac{1}{1 - \left(\frac{I_z}{I}\right)^2} \tag{2.30}$$

Taking the ambient temperature ϑ_u into account, the total temperature rise of the plant when hot becomes

$$\vartheta_g = \vartheta_u + \vartheta\left(1 - e^{-\frac{t}{T_t}}\right)$$

and the continuously permissible temperature ϑ_z is reached after the time t_g

$$t_g = T_t \ln \frac{1}{1 - \left(1 - \frac{\vartheta_u}{\vartheta_z}\right)\left(\frac{I_z}{I}\right)^2} \tag{2.31}$$

Figure 2.12a shows the temperature rise of electrical plant in relation to time with the overcurrent as parameter when neglecting the ambient temperature (equation (2.30)) and Fig. 2.12b the influence of the ambient temperature on the total temperature rise of the plant. It can be seen that the load current and the

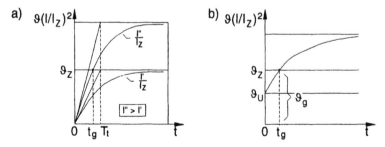

Fig. 2.12: Temperature rise of electrical plant in relation to time when neglecting a) and when taking account b) of the ambient temperature ϑ_u [2.19]

ambient temperature (cooling medium temperature) have a considerable influence on the limit time t_g.

2.6 Real power deficit

Under steady-state normal operating conditions, a balance exists between the generated power and the consumer load plus the losses of the power system itself. Should a generator unit or an important interconnecting line be lost due to a fault or a switching operation, there will be a shortage of real power in at least a part of the system. Fig. 2.13 shows an example of such a situation. The industrial network in this example has its own generators, but normally draws additional real power P_{ext} from the power grid to cover the total real power requirement P_g. Neglecting the power system losses, the following applies under normal load conditions

$$P_g = P_{ext} + P_{int} \tag{2.32}$$

Should for some reason the supply from the power grid be lost, there is a deficit of real power in the industrial power system which immediately initiates a reduction of turbo-alternator speed and thus also of system frequency. The reduction of frequency is influenced by the following factors:

- the value of the power deficit ΔP in relation to the total power of all the generators still in operation
- the resulting polar moment of inertia of all the rotating machines in the island power system. This is characterized by the constant of inertia H which is defined as the ratio between the kinetic energy and the apparent power of the rotating machines.
- the load reduction factor k_0, which defines the influence of frequency reduction on the real power requirement of the loads
- the type of turbine governor and the value of the spinning reserves

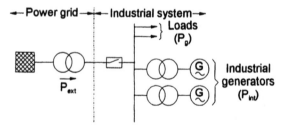

Fig. 2.13: An industrial power system with its own generators which imports additional power

The first two influencing factors determine the initial rate of frequency reduction, i.e. what is referred to as the frequency gradient at $t = 0$, to which the following applies [2.20]

$$\frac{df(0)}{dt} = -\frac{\Delta P}{2H} f_N \qquad (2.33)$$

where f_N is the rated frequency of the power system. Typical values for H are

- hydro-generators $H = (1.5 \dots 5.5)$ s
- turbo-alternators with HP and LP turbines $H = (2 \dots 10)$ s
- industrial turbo-alternator sets $H = (3 \dots 4)$ s

The higher values result for the units with the lower powers.

The remainder of the influencing factors given above, i.e. the load reduction factor, the type of turbine governor and the spinning reserves, determine the remainder of the frequency characteristic. If there are no spinning reserves and the turbo-alternator sets in operation are fully loaded, the loss of all the generators and complete collapse of the system can only be avoided by shedding sufficient load to reestablish a balance between generation and load. This function is performed by an automatic underfrequency load-shedding scheme.

2.7   Other faults and abnormal operating conditions

Apart from the relatively frequent faults and disturbances which afflict electrical power systems and their components given above, there are also other more infrequent faults or faults only concerning particular items of plant. The most important of these are:

- power swings in power grids
- underexcitation of synchronous generators
- overfluxing of power transformers
- asynchronous operation of synchronous motors
- mechanical defects (e.g. leaking power transformer oil tanks, linkage breaks on power transformer tap-changers, torsional stress on the shaft trains of turbo-alternator sets etc.)

These and other kinds of faults and their detection are discussed in more detail in Chapters 7 and 9.

3

The Main Criteria for Detecting Faults

3.1 Overcurrent

An overcurrent is when the maximum continuous load current permissible for an item of electrical plant is exceeded. Overcurrents are characteristic of phase and ground faults, but can also occur during normal operation, e.g. when energizing power transformers, induction motors etc. An overcurrent can thus be used as a criterion for detecting phase and ground faults, providing overcurrents of the kind mentioned above either cannot occur in the particular system, or are prevented from causing tripping by functions included in the protection (e.g. time delay, blocking by harmonic detectors, discrimination by pick-up setting etc.).

Because of its simplicity and for reasons of cost, the overcurrent criterion is frequently applied as overload protection, although in principle this is incorrect, since without some ancillary, an overcurrent detector does not take account of the load level prior to an overcurrent, which otherwise might be permissible at least for a short period. Correct overload protection which enables full advantage to be taken of the thermal capability of the plant necessitates the registration of the load level by means of a thermal image of the protected unit or some other form of storing the thermal data.

Overcurrent protection devices continuously monitor the current being conducted by the protected unit and issue a tripping command to the circuit-breaker when the current exceeds the setting (possibly for the period of a fixed time delay or a time delay determined by the operating characteristic) to isolate the respective item of plant from the power system.

In the case of time-delayed overcurrent protection, short peaks of current, which are higher than the pick-up setting but shorter than the time delay cannot cause tripping of the circuit-breaker. There is a difference, however, between the level at which a protection device picks up and the level at which it resets. This is particularly so with electromechanical protection devices, but also exists to a lesser extent on solid-state devices. The quotient of the two levels is termed the reset ratio. The following condition thus results for the setting of the pick-up current

$$I_{Fmin} > I_{pick-up} > \frac{I_{Lmax}}{r} \tag{3.1}$$

where

I_{Lmax} = maximum permissible load current of the protected unit
I_{Fmin} = minimum fault current at the relay location for a fault at the limit of the protected zone
$I_{pick-up}$ = relay (set) pick-up current
r = relay reset ratio (i.e. reset current / pick-up current)

To avoid any risk of the protection - especially instantaneous overcurrent protection devices - operating for a fault in a neighbouring zone, the setting must be chosen higher than the maximum fault current at the limit of the protected zone. This situation can arise, for example, for the three-phase fault current (initial value for instantaneous devices) downstream of a power transformer, or the three-phase fault current at the end of a MV transmission line etc. Absolute discrimination is scarcely possible with such methods, above all in view of the differing operating and power system conditions, and an impedance or comparison protection scheme becomes necessary.

The pick-up of the overcurrent protection in the case of an induction motor is normally set higher than the maximum AC starting current. The latter depends on the design of the motor and is usually between 3 and 6 times the rated current of the motor.

3.2 Differential (circulating) current

In this case, the amplitudes of the current upstream and downstream of the protected unit are compared. A fault is deemed to exist as soon as the difference between the two currents is no longer zero. For this reason, one of the names for this kind of protection scheme is differential protection; another is circulating

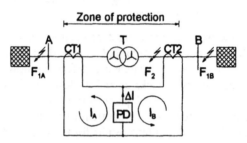

Fig. 3.1:   Example of fault detection using a differential current criterion

current, because as can be seen from the basic circuit for protecting a two-winding power transformer shown in Fig. 3.1, the current circulates around the circuit
formed by the c.t. secondaries CT1 and CT2 and the leads connecting them. The
differential or spill current

$$\Delta I = I_A - I_B \tag{3.2}$$

flowing through the protection device PD is monitored. Neglecting all inaccuracies (i.e. differences between the ratios of the two sets of c.t's and the ratio of the
power transformer, differences due to power transformer losses and any tap-
changer etc.), the differential current ΔI will equal zero under all normal operating conditions since $I_A = I_B$. This applies for each of the phases of the three-phase
system. The equation also applies, once again neglecting the above inaccuracies,
for phase and ground faults outside the zone of protection (fault locations F_1 and
F_2).

     In the event of a fault in the protected zone, the downstream current (e.g. I_B)
of the faulted phase will reverse and the protection device will measure the sum
instead of the difference of the currents, i.e.

$$\Delta I = I_A + I_B$$

If the power transformer is only supplied from one side, this reduces to

$$\Delta I = I_A$$

     In practice, however, a considerable differential current can sometimes occur
under normal load conditions or during a "through fault" even though there is no
fault in the protected zone. This is especially the case when a through fault is accompanied by c.t. saturation.

     Therefore the pick-up setting of the protection device has to be chosen to be
reliably below the current ΔI for all possible faults in the protected zone on the
one hand, but safely above all the differential (operating) currents during normal

operation and external faults caused by the inaccuracies and discrepancies given above on the other. Since this condition frequently cannot be satisfied in practice, the level of the current flowing through the protected unit is used as a stabilizing (restraining) influence and the protection measures the quotient of operating current divided by restraining current, i.e. the sensitivity of the protection reduces as the through current increases. This enables the protection to discriminate between a fault and a load condition in the majority of cases. Harmonic restraint is also used to maintain protection stability during the inrush currents which flow when energizing a power transformer and as a result of overvoltages.

High-speed protection devices also make it necessary to match the c.t's to the protected unit not only with respect to amplitude, but also with respect to their transient behaviour, i.e. especially the secondary time constants on both sides of the protected unit. On the other hand, they can also be used in conjunction with saturation detectors or even enable the entire measurement procedure to be completed before a c.t. has time to saturate.

The variations of power transformer ratio caused by the tap-changer can be accommodated by an adaptive protection system.

3.3 Difference of current phase-angles

The protection compares the phase-angles of the currents flowing into and out of the protected unit and is therefore also referred to as phase comparison protection. Whereas a differential or circulating current scheme determines the difference between the amplitudes on both sides of the protected unit, in this case it is the difference between the phase-angles which is of consequence (Fig. 3.2a). Under normal load conditions and during an external fault (F_{1A}, F_{1B}), the phase difference

$$\Delta\varphi = \varphi_A - \varphi_B \tag{3.3}$$

between upstream and downstream currents is 180°, but during an internal fault on the protected unit $\Delta\varphi \approx 0$.

The capacitances of the conductors give rise to a degree of phase-shift between the currents I_A and I_B and this makes a safety margin necessary to avoid any risk of false tripping. Figure 3.2b shows the operating and restraint areas for a phase comparison protection scheme. The restraint area extends on both sides of the 180° mark by $\pm(30...60)°$.

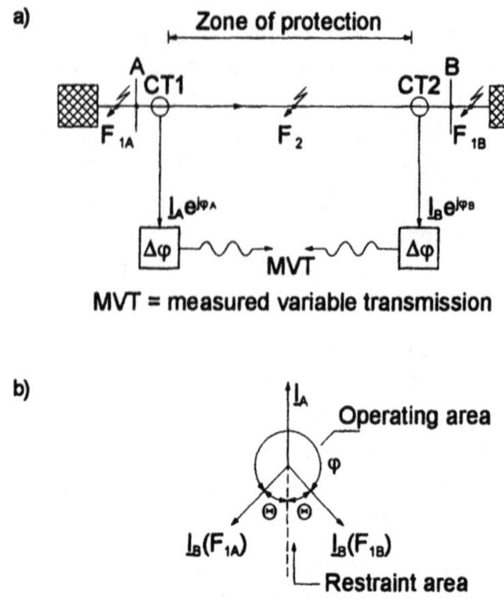

Fig. 3.2:   Example of fault detection by monitoring the difference between the phase-angles of the currents on both sides of the protected unit a) and the corresponding vector diagram b)

3.4   Over- and undervoltage

In a three-phase power system, the phase-to-neutral and phase-to-phase voltages at the load are influenced by the voltage drop along the line and therefore by the load itself, but may only vary within given limits. Should the voltages lie outside these limits, an abnormal operating condition or a fault is indicated. Overvoltages can be the result of defective voltage regulators on generators or power transformers, or of load shedding, or poor power factor regulation. Atmospheric overvoltages (lighting strikes, travelling waves etc.) do not belong to this group, because they are usually taken care of by lighting arresters. Undervoltage is mostly a consequence of a fault.

A voltage criterion is sometimes combined with another criterion, for example, overvoltage with overfrequency on generator step-up transformers and undervoltage with overcurrent on small generators. A better alternative in the latter case is an underimpedance scheme.

3.5 Power direction

A directional unit is used in combination with an overcurrent unit at locations on the power system where the overcurrent criterion on its own is insufficient to preserve discrimination. This principally concerns overcurrent units on ring lines, parallel circuits or power transformers. Directional units are also used on turbo-alternator sets to detect abnormal operating conditions, i.e. dangerous motoring of the generator after the main steam valve has been closed and asynchronous operation of synchronous motors. The real power reverses direction in the former case and the apparent power in the latter.

Figure 3.3a shows the typical application of a double circuit line, where the overcurrent units are only able to discriminate between faults inside and outside the zone of protection with the aid of the directional units. Since for a fault on just one of the circuits (e.g. F_1 on line 1) the fault is also supplied via the healthy line and the remote station, overcurrent protection on its own would also trip the healthy line (circuit-breaker 4 on line 2). Discrimination is achieved by adding directional units at locations 2 and 4 so that only the circuit-breakers of those lines are tripped which are conducting fault energy away from the busbar. This applies to circuit-breaker 2 for F_1, but not to circuit-breaker 4. Neither circuit-breaker 2 nor 4 are tripped for the external fault F_3.

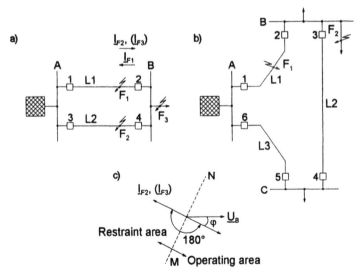

Fig. 3.3: Application of a power direction criterion for phase and ground fault protection
a) double circuit line   b) ring line   c) vector diagram

In an AC system a reference is needed to determine power direction and for this purpose the busbar voltage is used. Since this reference voltage may also fail should a fault occur on or near the busbar, either a memory feature (tuned circuit) must be included to ensure that sufficient voltage is available for reliable measurement, or the scheme must be designed such that tripping is defined in the event of complete voltage collapse and loss of the reference, although it may not necessarily be selective.

Figure 3.3c shows the operating and restraint areas of the directional characteristic. Note the phase-angle of the fault current in this diagram.

The same considerations apply to the ring system of Fig. 3.3b.

3.6 Symmetrical components of current and voltage

Often the symmetrical components of the phase currents and voltages are more suitable for protection purposes than the phase values themselves. Typical examples are the detection of imbalances by monitoring the level of the negative-sequence component and the detection of positive and negative-sequence components of the current in comparison protection schemes to reduce the volume of data being transmitted between the stations. In these applications, filters are used to extract the symmetrical components from the three-phase current and voltage systems.

The principal kinds of faults which are detected by monitoring the symmetrical components are:

- phase-to-phase and ground faults in solidly and low-resistance grounded systems
- ground faults in ungrounded or inductively (Petersen coil) grounded systems
- asymmetric system configuration, asymmetric load and open-circuit phase conductors

Depending on the problem to be solved, the protection may evaluate the amplitudes of the symmetrical components or the phase relationship between them.

3.7 Impedance

Impedance is the criterion measured to detect faults on transmission systems or underexcitation or out-of-step conditions on generators.

The detection of faults on transmission systems is based on the fact that the impedance measured at the relay location under normal load conditions (i.e. load impedance = quotient of load voltage divided by load current) is normally dis-

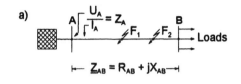

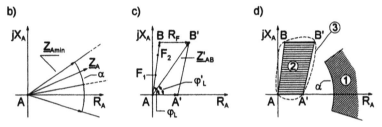

Fig. 3.4:   Use of the impedance criterion to detect phase and ground faults
a) power system configuration
b) impedance under load conditions
c) impedance measurement under fault conditions
d) impedance areas: ① load   ② fault   ③ relay characteristic

tinctly higher than the impedance of a fault. Also under certain conditions, the impedance of the fault loop from the relay location to the fault is proportional to the distance between the two. Figure 3.4 illustrates the behaviour of the impedance measured during normal operation in relation to the impedance measured during a fault.

It has become conventional to show the relationships in the R/X plane, i.e. a vector diagram (voltage divided by current) in which the origin represents the relay location and the beginning of the protected line, the ordinate the reactance X and the abscissa the resistance R. Under load conditions (Fig. 3.4b), the absolute value of the impedance measured (for a constant voltage U_A) only depends on the line current I_A, while the argument of the impedance depends on the power factor at which the electrical power is transported to the load. The minimum load impedance Z_{Amin} is reached at the maximum load current. The impedance vector lies on the straight line AB representing the transmission line for phase and ground faults (Fig. 3.4c) which do not include any resistance at the fault location. The inclination of the straight line AB is determined by the impedance angle $\varphi_1 = \arctan(X_{AB} / R_{AB})$ of the transmission line. If a fault resistance R_F exists at the fault location (e.g. due to arc resistance), a higher impedance is measured, but the impedance angle of the line is reduced. This condition is shown for a fault close to the B end of the line.

The operating characteristic of a protection device can be represented in the R/X diagram by a closed curve. Tripping normally takes place when the measured value of the impedance lies within the curve and does not take place for all other positions of the impedance vector.

By marking the prospective areas described by the load impedance ① and the fault impedance ② in an R/X diagram as shown in Fig. 3.4d, it can be seen that the operating characteristic ③ of the protection device must be positioned such that all faults are detected, i.e. that the area ② is fully enclosed, but it must never encroach on the load area ①. The art of protection lies in being able to device a characteristic and a setting for the protection device, which are able to fulfil this requirement under all power system operating conditions.

It is a relatively simple matter to make an impedance unit dependent on direction; such units are used in directional distance relays.

The application of an impedance measurement for the detection of out-of-step and underexcitation conditions is dealt with in Chapter 8.

3.8 Frequency

The deviation of the frequency of an electrical power system from its rated value is an indication of an imbalance between real power generation and load demand, too little power causing the frequency to fall and too much power causing it to rise. Since the power system also supplies synchronous clocks, it is important for the frequency to remain constant within narrow limits. Frequency deviations are thus relatively small and fluctuations in the mHz/MW range for a given step change of real power is a measure for the strength and stability of a power system. Thus the systems used for measuring frequency have to be very accurate. In the case of a frequency reduction due to a deficit of real power, it is standard practice to shed load in steps until a balance between generation and demand has been re-established and the frequency restored. Where the deficit is large, a faster load shedding response is achieved by varying the size of the blocks of load shed in relation to the rate-of-change of frequency (frequency gradient) df/dt. On no account, however, may the rate-of-change be used on its own for load shedding, i.e. without being enabled by a unit measuring the absolute frequency, because similar frequency gradients can occur during normal operation when connecting and disconnecting parts of the system [3.4, 3.5].

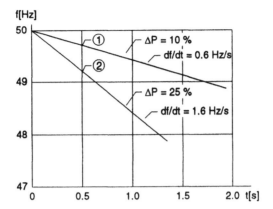

Fig. 3.5:   Reduction of power system frequency in relation to time for different levels
of power deficit

Figure 3.5 shows the reduction in frequency in relation to time for real power
deficits of 10 % (curve ①) and about 25 % (curve ②). The corresponding gradients are 0.6 Hz/s and 1.6 Hz/s.

3.9   Other criteria monitored by protective devices

There are other criteria which are monitored for protection purposes and these
are either described together with the associated applications in Part B or are
self-explanatory. Such criteria are:

- temperature as overload criterion (typical application: transformer oil temperature). If the temperature is not measured directly, but modelled in a thermal image of the protected unit conducting the load current, the corresponding devices are described in Section 2.5.
- rate of oil flow, accumulation of gas (for detecting internal transformer faults)
- harmonics in the neutral current and/or neutral voltage for detecting ground
faults in inductively grounded systems
- harmonics in the generator current for detecting internal generator faults
- transient current and voltage signals, travelling waves etc., on transmission
lines for detecting faults

3.10 Summary

The principal criteria for detecting faults and abnormal operating conditions on electrical power systems are listed in Table 3.1. The practical application of these criteria is the subject of Parts B and C.

Table 3.1:   The principal kinds of power system faults and system variables (criteria) used to detect them

No.	Kind of fault	Variable used for detection
1	Phase faults in general	Phase current $\quad$ I Current difference $\quad \Delta I$ Difference of current $\quad \Delta \varphi$ phase-angle Phase voltage $\quad$ U Power direction $\quad P\rightarrow$ Impedance $\quad$ Z
2	Asymmetric faults (ground, phase-to-phase and phase-to-phase-to-ground)	As above plus: Negative and zero-sequence components of current (I_2, I_0), voltage (U_2, U_0) and the power directions (P_2, P_0) $\rightarrow$
3	Ground faults	Zero-sequence components of: - current $\quad I_0$ - voltage $\quad U_0$ - power direction $\quad P_0\rightarrow$
4	Overload	Phase current $\quad$ I Temperature $\quad \vartheta$
5	Asymmetric configuration Asymmetric load Interrupted conductor	Negative-sequence $\quad I_2$ component
6	Real power deficit	Frequency $\quad$ f Rate-of-change $\quad$ df/dt of frequency
7	Real power excess	Frequency $\quad$ f

4

Instrument Transformers for Protection Purposes

4.1 Current transformers (c.t's)

4.1.1 Conventional c.t's

C.t's step down the primary system currents of up to a few thousand amperes to a level which the measurement and protection circuits can handle, usually 1 or 5 A at rated primary current. In HV installations, the primary windings are at the potential of the HV system and therefore the c.t's also have to provide the insulation between the two potentials. Conventional c.t's are those with a closed iron core.

Depending on the purpose for which they are intended, the cores of conventional c.t's differ in design, a distinction being made between measurement, metering and protection cores. The difference mainly concerns their behaviour when conducting overcurrents. While measurement and metering cores are deliberately designed not to supply any secondary current above 1.2 times rated current to avoid overloading the sensitive instruments connected to them, it is just in the overcurrent range that the protection cores have to transform the primary current correctly to enable the protection devices to accurately detect a fault. In their range of operation, measurement and metering cores, are also required to transform current amplitudes and phase-angles faithfully and with much greater accuracy. Similarly low phase errors are required of protection cores even in the

39

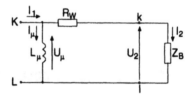

Fig. 4.1:   Simplified equivalent circuit of a conventional protection c.t.
I_1 = primary current, I_2 = secondary current, I_μ = magnetizing current, R_w = secondary winding resistance, L_μ = main inductance, Z_B = impedance of the burden including leads

overcurrent range when supplying directional protection devices. For some kinds of protection, the response of the c.t. cores to transients, harmonics and/or DC components is of consequence. The majority of these requirements are regulated by standards and the c.t's accuracy classes derived from them. The remainder of this chapter is only concerned with the protection cores of the c.t's.

Taking account of the various parameters of a c.t., the equivalent circuit can be simplified as shown in Fig. 4.1 [4.1].

The main parameters of a c.t. of consequence for protection applications are:

- rated currents of primary and secondary
- Class (ratio and phase errors, measuring range)
- rated burden
- rated overcurrent factor (knee-point voltage)

There are standardized series of values for the primary rated current I_{1N}. According to the international standard IEC 187, the "major series" is 10, 15, 20, 30, 50 and 75 A and its decimal multiples and quotients and the "minor series" 12.5, 25, 40 and 60 A and also the corresponding decimal multiples and quotients.

Generally the primary rated current will be chosen somewhat higher than or equal to the continuously permissible load current of the item of plant to be protected (exception: c.t's designed for continuous overload which are referred to as wide-range c.t's).

The most common secondary rated currents I_{2N} are 1 A and 5 A. A secondary current of 5 A is usual in MV systems and standard practice in the USA. In Europe, 1 A is used at all system voltages above 110 kV to minimize the burden of the leads.

The rated ratio of a c.t. is

$$K_N = \frac{I_{1N}}{I_{2N}}$$

5P and 10P are the accuracy classes used mainly for protection cores, P standing for protection and the number signifying the total percentage error F_g defined for the entire overcurrent range. The following applies

$$F_g = \frac{1}{I_1} \sqrt{\frac{1}{T} \int_0^T (K_N i_2 - i_1)^2 \, dt} \quad 100\% \tag{4.1}$$

where

i_1, i_2 = instantaneous values of primary and secondary currents in [A]
I_1 = rms value of the primary current in [A]
T = period of the rated system frequency in [s]
K_N = rated ratio of the c.t.

Class 10P c.t's with a total error F_g of 10 % are normally only used for simple overcurrent relays; Class 5P is chosen for all other protection devices.

The rated power S_N of a c.t. is defined as its apparent power in VA at rated secondary current and burden. This power may not be exceeded in operation, i.e. the sum of the burdens of all the devices connected to the c.t. plus the burden of the leads must not be greater than S_N. The preferred rated powers for protection cores are 10, 15, 30 and 60 VA. Where a solid-state relay is connected to a c.t., its burden is so low that the load on the c.t. is determined mainly by the leads.

The rated overcurrent factor n_N expresses the overcurrent response of the c.t. cores. It is the multiple of the primary rated current I_{1N} at which the total error F_g at rated c.t. burden does not exceed 5 % for a Class 5P c.t., respectively 10 % for a Class 10P c.t. This definition is illustrated for a Class 5P20 and a Class 10P15 protection core in Fig. 4.2. The classification in the case of the first core means that at 20 time rated current the error is less than 5 % and in the case of the second core, that at 15 times rated current the error is less than 10 %. Typical values for n_N of protection cores are 10, 15, 20 and 30.

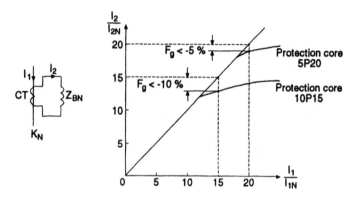

Fig. 4.2: Graphical representation of the rated overcurrent factor n_N for two classes of c.t's

Where the actual secondary burden of a c.t. differs from the rated burden, the effective overcurrent factor n_t varies according to the equation

$$n_t = n_N \frac{S_N + S_E}{S_t + S_E} \tag{4.2}$$

where

n_t = effective overcurrent factor
n_N = rated overcurrent factor
S_N = rated c.t. power in [VA]
S_t = actual secondary burden (including lead burden) in [VA]
S_E = burden of the c.t. itself in [VA] (approx. $0.1 S_N$)

It follows from equation (4.2) that the effective overcurrent factor n_t can be increased by reducing the burden S_t. This fact is frequently used in practice to improve c.t. overcurrent performance.

According to equation (4.3), the overcurrent factor of the core must be chosen such that the operation of the protection is not impaired by the core entering saturation. The maximum rms value of primary current which can be measured without exceeding the values for the total error given above is taken as reference current. Thus

$$n_g = \frac{K_s \, I_{ref}}{I_{tN}} \tag{4.3}$$

where

n_g = required overcurrent factor

K_S = safety margin chosen between 1.2 and 2 depending on the operating time t_a of the protection (lower values for long operating times, i.e. greater than 0.3 s, and higher values for short operating times)

I_{ref} = reference current depending on the kind of protection

In the case of simple overcurrent protection devices, the pick-up current of the protection referred to the primary is used as reference current. For distance protection, the initial value of the AC fault current for a three-phase fault at the limit of the first zone is usually taken as reference current. Differential protection schemes must be designed such that false tripping due to c.t. saturation cannot occur at the steady-state through fault current. For this reason, the maximum initial AC through fault current is inserted for I_{ref} in equation (4.3).

It is possible for very fast protection devices to complete their measurement before the c.t. enters saturation. Assuming the above worst-case conditions, this necessitates, however, determining the time from the inception of the fault until the core saturates, which must be greater than the operating time of the protection [4.15].

There are some protection devices available especially for bus zone protection, which tolerate c.t. saturation before they have completed their measurement. One such protection is a high-impedance scheme, for which the c.t's have to be specifically designed and normally have relatively large core cross-sections.

Modern measuring principles also permit the protection to detect the start of saturation and thus enable precautions to be taken to prevent the saturation from having any adverse effects [4.17, 4.19].

C.t's are usually constructed with several cores. One or more are intended for measurement and metering, but those protection devices are connected to them which have to measure currents accurately at fractions of the rated current. Examples of these are reverse power protection and underexcitation protection of synchronous generators.

In the case of digital control and protection units for feeders, protection and metering functions are supplied by the same core so that for this type of equipment the protection core must be sufficiently accurate at low currents.

If because of the importance of the plant two sets of protection are installed for greater protection reliability, it is logical for the two sets to be connected to independent c.t. cores. Figure 4.3 shows the example of a c.t. in one of the phases of a large generator rated at 1000 MVA having a single primary and four cores with one secondary winding each.

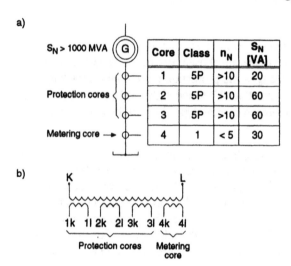

Fig. 4.3:   Example of a c.t. with four independent cores
             a) circuit diagram and core specifications   b) symbol

Table 4.1 shows c.t. circuits for obtaining the phase, neutral and differential currents. The practical application of these and other arrangements are discussed in detail in Part B.

Transient behaviour of conventional c.t's

In power systems with a relatively large value of L_N/R_N for the fault loop, the transient behaviour of the c.t's during the operating time of the protection must be taken into account. Large values of L_N/R_N indicate long time constants for the DC component contained in the fault current and this is especially significant close to large generators where the DC time constant of the fault current can reach 400 ms. DC components can cause c.t's with closed iron cores to saturate at relatively low current amplitudes. The primary ampere-turns for the DC component are not compensated by back e.m.f. ampere-turns on the secondary side,

Table 4.1: Different methods of connecting c.t's for protection purposes

No.	Circuit designation / Measured variable	Circuit
1	Y connection / Phase current	
2	V connection / Phase current	
3	Holmgren / Neutral current	
4	Delta connection / Phase-to-phase	
5	Longitudinal differential / Differential current	

with closed iron cores is generally (depending on the burden) much longer than the primary time constant and therefore almost the entire DC component contributes to the flux of the core.

Assuming a fully off-set fault current, i.e.

$$i_1 = \hat{I}_{1F}\left(e^{-\frac{t}{T_N}} - \cos\omega t\right)$$ (4.4)

where

$\hat{I}_{1F}$ = peak value of the steady-state fault current
T_N = L_N/R_N = power system time constant
ω = angular velocity

the transient behaviour of the c.t. can be determined from the equivalent circuit according to Fig. 4.1. Assuming further that the resistance of the secondary winding is added to the resistive portion of the burden, the following differential equation is obtained

$$L_\mu\frac{di_\mu}{dt} = R'_B\, i_2 + L'_B\frac{di_2}{dt}$$

and taking account of the fact that $i_2 = i_1 - i_\mu$,

$$\frac{di_\mu}{dt} + \frac{1}{T_W}i_\mu = q\frac{di_1}{dt} + \frac{1}{T_W}i_1$$

where

T_W $= \dfrac{L_\mu + L'_\mu}{R'_B}$ = c.t. time constant including burden

R'_B $= R_B + R_W$

q $= \dfrac{L_B}{L_\mu + L_B}$

Since the flux of the core varies in proportion to the magnetizing current, equation (4.5) gives the excursion of the current in relation to time for the fully off-set fault current obtained from equation (4.4). This takes account of the permissible simplification that $(\omega T_W)^2 \gg 1$ and $\omega T_N \gg 1/(\omega T_N)$ and also the substantiated assumption that $L_B = 0$. The following results

$$i_\mu = \hat{I}_{1F}\left[\frac{T_N}{T_N - T_W}\left(e^{-\frac{t}{T_N}} - e^{-\frac{t}{T_W}}\right) - \frac{1}{\omega T_W}\sin\omega t\right] \tag{4.6}$$

$\underbrace{}$ $\underbrace{}$

DC comp.   AC comp.

Since $B = K_W i_\mu$, K_W being a constant for the core determined by its permeability and length, the excursion of the induction is directly given by the above equation (4.6). Figure 4.4a shows the corresponding partial fluxes and the total flux assuming that the latter is less than the saturation flux B_S. It can be seen that the flux B_{DC} produced by the DC component is much greater than that of the AC component B_{AC}. The same conclusion results applying the usual time constants from the calculation of the quotient "DC component/AC component" in equation (4.6). The power system time constant obtained for a three-phase fault ranges from 50 to 300 ms and, as mentioned above, in extreme cases can be as high as 400 ms. With normal burdens, secondary circuits of protection c.t's have time constants of a few seconds.

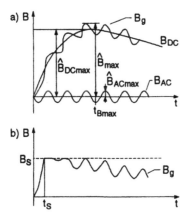

Fig. 4.4:   The excursion of the flux of a c.t. for a fully off-set fault current
a) without saturating ($B_{max} < B_S$)   b) with saturation ($B_{max} > B_S$)

According to Fig. 4.4a, the total flux reaches a maximum when its DC component B_{DCmax} is a maximum, i.e. at the instant the first summand in equation (4.6) becomes a maximum. This time is given by

$$t_{Bmax} = \frac{T_N\, T_W}{T_W - T_N} \cdot \ln \frac{T_W}{T_N} \tag{4.7}$$

Under the condition that $\omega T_W \gg 1$, equation (4.6) can be used to calculate the maximum possible flux for the time t_{Bmax}, the amplitude of the AC component being determined for $\sin \omega t$. From this the ratio $B_{max}/B_{AC} = K_t$ becomes

$$K_t = \frac{B_{max}}{B_{AC}} = \omega T_N F_L + 1 \tag{4.8}$$

Where K_t is the "transient" or "oversizing" factor and F_L is

$$F_L = \left(\frac{T_N}{T_W}\right)^{\frac{T_N}{T_W - T_N}} \tag{4.9}$$

For an ideal c.t., i.e. for $L_\mu = \infty$ (and therefore also $T_W = \infty$), $K_t = \omega T_W + 1$; Fig. 4.5 gives the values of the oversizing factor for a practical c.t. for which $L_\mu \neq \infty$. It can be seen that even for relatively low power system time constants T_N, the oversizing factor K_t can assume significant values. For example, K_t becomes approximately 25 for $T_N = 100$ ms and $T_W = 1$ s. This means that a c.t. with an iron core 25 times larger than normal would have to be installed to preclude saturation. C.t. costs become even worse, if the remanence factor K_r of the core, which is defined as the quotient B_r/B_S (B_r = residual flux), is taken into account. The condition for avoiding saturation is then expressed by

$$\xi > \frac{K_t}{1 - K_r} = \frac{\omega T_N F + 1}{1 - K_r} \tag{4.10}$$

$K_r = 0.8$ for c.t's with closed iron cores, i.e. K_t is increased by a factor of 5 which corresponds to an enormous cross-section of the iron. This is technically and economically difficult to achieve so that in practice c.t's are usually designed not to saturate just for the operating time t_a of the protection. In this case, the following applies according to [4.1]

$$t_s = \frac{1}{3} t_{Bmax} \ln\left[1 - \frac{0.8 n_t (1 - 1.5 K_r)}{K_t - 1} \cdot \frac{I_{tN}}{I_t}\right] > t_a \tag{4.11}$$

where t_s is the time between the inception of the fault and the start of saturation.

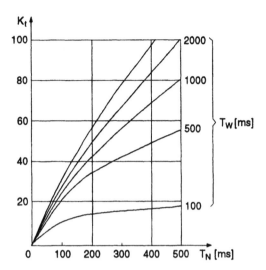

Fig. 4.5: Oversizing factor K_t for c.t's with different time constants T_W in relation to the power system time constant T_N

From equations (4.10) and (4.11) it is obvious that in many cases, economically acceptable conventional c.t's cannot be prevented from saturating, especially where fault currents are exceptionally high and operating times long (back-up protection). Should saturation occur, the secondary current is no longer a faithful representation of the primary current (Fig. 4.6), the ratio and phase errors become inadmissibly high and therefore discriminative operation of graded protec-

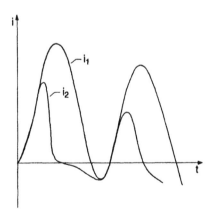

Fig. 4.6: Secondary current of a c.t. saturated by a fully off-set primary current

tion devices is no longer certain [4.3, 4.4, 4.7]. Secondary currents distorted by saturation have a high harmonic content, which also can influence the operation of protection devices [4.6, 4.10]. This is a further reason for the development of the protection devices mentioned previously, which can also operate discriminatively with distorted secondary currents or operate before saturation takes place [4.15]. Classes of c.t's have also been introduced for special kinds of protection with the type designations TPX, TPY and TPZ (TP = transient performance).

Class TPX includes oversized c.t's with closed iron cores which are designed according to the principles described above. Their main disadvantage is the large cross-section of the iron and the resulting costs and installation problems [4.7] and the extremely high value of the residual flux factor K_r.

4.1.2 Air-gap c.t's

C.t's with an air-gap in the magnetic path have the type designations TPY and TPZ. The introduction of an air-gap in the iron core reduces the main inductance L_μ and therefore also the c.t. time constant T_W. In consequence, a much smaller oversizing factor K_t is permissible (see Fig. 4.5), which means that the cross-section of the iron is also greatly reduced in comparison with the Class TPX.

Classes TPY and TPZ differ in the number and length of the air-gaps and therefore in the time constants and residual flux factors.

Class TPY c.t's, which are also referred to as anti-remanence air-gap c.t's have only a few (usually two) small air-gaps to limit the residual flux after a fault has be tripped. The residual flux factor is about 0.1 and the c.t. time constant about 0.1 to 1 s. If after a fault has been tripped the load current is not restored, a c.t. with a closed iron core can maintain its residual flux for hours or even days and depending on the half-cycle when the line or plant is reenergised can immediately saturate. Anti-remanence air-gap c.t's are therefore recommended in powerful systems.

Class TPZ c.t's are referred to as linearized c.t's and have the best transient performance. Their iron cores have several large air-gaps, which reduce the residual flux to almost zero and the time constant to less than 60 ms. A reduction of the time constant below this figure is not permissible, because the phase error is already 3° and would otherwise become greater. The cross-section of the core is considerably less than for TPX or TPY c.t's. The main disadvantages of linearized c.t's are the imprecise transformation of the DC component in the fault current, which is inadmissible for certain kinds of protection, and the slow decay of the secondary current (DC portion) after the primary current has been interrupted. The latter property prohibits the use of Class TPZ cores for breaker back-up protection (see Section 7.3.6).

The ratio error for the DC component is expressed by the first term of equation (4.6). It reaches a maximum identical to the value of F in equation (4.9) at the time t_{Bmax} as given by equation (4.7), i.e.

$$\varepsilon_{max} = \left(\frac{I_{1DC}}{K_N \, I_{2DC}}\right)_{t_{Bmax}} = F_L \quad 100\% \quad (4.12)$$

where I_{1DC} and I_{2DC} are the maximum values of the primary and secondary DC components. Figure 4.7 shows the absolute value of ε_{max} in relation to the ratio T_N/T_W. Assuming a time constant of about 60 ms for TPZ c.t's and typical time constants of $T_N = 100 \ ... \ 500$ ms for the power system, a range of $\varepsilon_{max} = 50 \ ... \ 75 \%$ is obtained for the ratio error, i.e. the value of the DC component in the secondary current is up to 75 % less than the DC component in the primary current [4.10].

Application of air-gap c.t's

The protection devices for lines in strong power systems should always be equipped with anti-remanence air-gap c.t's (Class TPY), because there is a high likelihood of c.t's without an air-gap saturating when allowing for the operating time of the back-up protection.

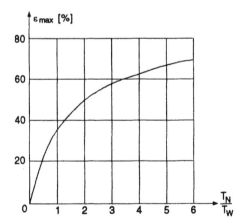

Fig. 4.7:   Maximum values of the ratio error ε_{max} between primary and secondary DC components

Table 4.2:   The most important parameters and characteristics of protection c.t's
Classes TPX, TPY and TPZ

Parameters and characteristics	C.t. Classes		
	TPX	**TPY**	**TPZ**
Ratio error F_i at rated current	± 1 %	± 1 %	± 1 %
Phase error δ_i	± 60'	± 60'	180' ± 20'
Remanence factor K_r	≈ 0.8	≈ 0.1	= 0
Time constant T_W	some s	0.1 ... 1 s	60 ± 6 s
Transformation of DC component	accurate	relatively accurate	poor

Class TPZ linearized c.t's are used mainly in differential protection schemes for high-power plant and for certain kinds of busbar protection [4.7]. When connecting two linearized c.t's to form a circulating current circuit for a differential protection scheme, the effective secondary time constants should be matched as far as possible. This is the reason why it is difficult to mix linearized and closed iron-core c.t's in a differential scheme.

4.1.3 Non-conventional c.t's

C.t's not conforming to the principles described above are referred to as non-conventional c.t's. These may, however, include interposing c.t's which do operate according to the principles of conventional c.t's and to which all the preceding comments apply. The application of non-conventional c.t's appears especially appropriate in EHV systems to avoid the expense of reaching the required insulation level with conventional c.t's.

Of the many non-conventional c.t's which have been successfully tested on EHV systems [4.12, 4.16], only the two most likely to prevail in the future are described below.

A *c.t. based on an opto-coupler* is shown in Fig. 4.8. The primary current I_1 is measured by a small conventional c.t. CT1 and applied to the modulator (1), which transforms it to a proportional voltage. The proportional voltage is then converted into either discrete impulses of variable duration (PWM = pulse width modulation) or a pulsed signal of variable frequency (PFM = pulse frequency modulation). The pulse width, respectively frequency of the resulting signal is proportional to the instantaneous value of the primary current. Following amplification in (2), this signal drives the infrared light-emitting diode (3), which forms

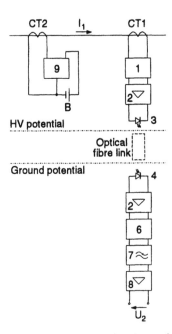

Fig. 4.8: Basic arrangement of a non-conventional c.t. which uses an opto-coupler as insulation barrier [4.1]

together with the photo-diode (4) at ground potential what is referred to as an opto-coupler. The light-emitting diode (3) — usually of gallium-arsenic — transmits light at a wavelength of 900 nm via the optical fibre link to the photo-diode, which converts the impulses of light back to proportional voltage impulses before they enter the demodulator (6) for conversion to a sinusoidal voltage. The modulation/demodulation process generates parasitic HF oscillations and these are eliminated by the low-pass filter (7). The resulting voltage signal now corresponds to the waveform of the primary voltage and is proportional to its amplitude and it only remains for it to be amplified by the amplifier (8) to obtain the powerful signal U2. Modules (1) and (2) which are at EHV potential require an auxiliary supply and this is provided by a battery charged by the c.t. CT2.

The principle of a *magnetic/optical Faraday* **c.t.** is shown in Fig. 4.9. The beam of light emitted by the light source (1) is conducted by optical fibres to the polarizer (2), the output of which is a beam of polarized light, i.e. light oscillating in only one axis. The polarized beam is deflected by a prism onto the quartz crystal (3A), in which a magnetic field is induced by the primary current I flowing in the coil. The magnetic field has the effect of rotating the polarization plane of the light, the angle of the rotation being given by the equation

$$\delta_p = V \cdot H \cdot l \qquad\qquad (4.13)$$

where

V   = Verdet constant; for quartz $5.23 \cdot 10^{-6}$ rad/Am
H   = magnetic field strength in the quartz crystal [A]
l   = length of the crystal rod in [m]

The beam of light is then deflected onto the quartz crystal (3B) at ground potential, which is in a magnetic field induced by the secondary current I_2. The direction of the field is chosen such that the polarization plane is compensated so that $I_1 W_1 = I_2 W_2$, W_1 and W_2 being the number of turns of the two coils around the quartz crystals. In consequence, a rotation δ_1 of the light in crystal (3A) is compensated by a rotation $\delta_2 = -\delta_1$ in crystal (3B). The levels of light in the two photo-detectors (A) and (B) are therefore equal and the input signal of the differential amplifier zero. A rotation of the polarization plane now causes one of the

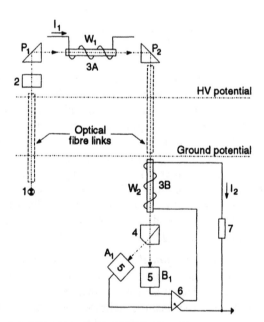

Fig. 4.9:   Principle of a magnetic/optical Faraday c.t.

two detectors to be illuminated more strongly than the other and thus a differential signal to be generated. The output signal of the differential amplifier is the secondary current I_2, which is a faithful representation of the primary current I_1. A secondary current only flows when the photo-detectors produce a differential signal, i.e. only for $I_1W_1 \neq I_2W_2$.

The advantage of this principle is that it does not require an auxiliary supply at EHV potential. Its disadvantages are that it is sensitive to mechanical vibration and has to be manufactured to high standards of precision.

4.2 Voltage transformers (v.t's)

4.2.1 Inductive v.t's

V.t's step down the primary system voltage without ratio or phase errors to a level which the measurement and protection circuits can handle. They also provide the insulation between the high potential of the primary system and the devices connected to the secondaries.

In principle, inductive v.t's are single-phase transformers running off-load. The corresponding equivalent circuit is given in Fig. 4.10.

A distinction is also made between the accuracy classes of v.t's for metering and protection applications. The designations for protection v.t's are 3P and 6P, the number signifying the percentage error F_U, which is defined as

$$F_U = \left(\frac{K_N U_2 - U_1}{U_1}\right) \ 100\% \tag{4.14}$$

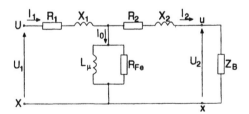

Fig. 4.10:   Equivalent circuit of an inductive v.t.

where

$K_N = \dfrac{U_N}{U_{2N}}$ = rated ratio of the v.t.

U_1, U_1 = rms value of the primary and secondary voltages

The phase error δ_U defined as the phase-shift between the voltage vectors U_1 and U_2 may not exceed $\pm 120'$ for Class 3P or $\pm 240'$ for Class 6P v.t's.

The maximum ratio and phase errors must remain within the range $0.05 U_{1N}$ to $K_S U_{1N}$, where K_S is the overvoltage factor of 1.5 for v.t's in solidly grounded or low-resistance grounded systems, respectively 1.9 in all other power systems.

The primary rated voltage is normally the rated voltage of the power system (i.e. for phase-to-phase connected v.t's which usually only applies to MV systems) or $1/\sqrt{3}$ times the power system voltage for phase-to-ground connected v.t's. The standardized secondary voltages are 100, 110, 200 and 220 V, respectively the corresponding values times $1/\sqrt{3}$. The rated voltage of secondary windings designed for connection in a 'broken-delta" arrangement to measure the neutral voltage is a third of the above values, i.e. 100/3 V.

The rated power of a v.t. is its apparent power in VA at rated secondary voltage and burden. Typical values are 25, 30, 50, 75, 100 and 200 VA.

The most frequently used v.t. circuits for supplying protection devices are listed in Table 4.3. Many v.t's have two secondary windings, one for measuring the phase-to-ground or phase-to-phase voltage, the other for measuring the neutral voltage during ground faults. The circuit of a v.t. of this kind is shown in Fig. 4.11.

The dynamic behaviour of inductive v.t's does not normally present any problems and does not have an adverse influence on the operation of the protection devices connected to them [4.1, 4.10]. Step changes of primary voltage (e.g. due to a close fault) are transferred to the secondary without appreciable displacement of the secondary voltage in relation to time. The waveforms of exponentially decaying DC components are also transferred faithfully to the secondary. Harmonics and HF oscillations are transformed truly within a wide band of frequencies. In certain situations, resonance phenomena can occur, where the frequency of a harmonic coincides with the resonant frequency of the v.t. [4.13].

Table 4.3: Different methods of connecting v.t's for protection purposes

No.	Circuit designation / Measured variable	Circuit
1	3 separate, single-phase, Y connected — Y and delta voltages	
2	2 separate, two-phase, V connected — Delta voltages	
3	3 separate, single-phase with 2 secondary windings — Y and delta voltages Neutral voltage	
4	1 single-phase — Neutral off-set voltage	

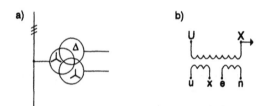

Fig. 4.11:   Single-line diagram a) and symbol b) of an inductive v.t. with two
              secondary windings

4.2.2   Capacitive voltage transformers (CVT's)

CVT's are also used as well as inductive c.t's at power system voltages higher than 110 kV, because they are cheaper for the higher voltages and have the added advantage of also being able to use the capacitive voltage divider needed for measurement as a coupling capacitor for PLC (power line carrier) communication.

A CVT is basically a capacitive voltage divider (C_1, C_2 etc.) with an inductive v.t. (IVT) tapping off the voltage across the lowest capacitor in the chain (i.e. the one with one leg connected to ground) to provide DC isolation. An inductance (L_p) compensates the phase error produced by the capacitive divider. Figure 4.12a shows the basic circuit and Fig. 4.12b a simplified equivalent circuit with a useful burden Z_B and a damping impedance Z_D, the latter to suppress ferro-resonance and transient balancing phenomena. The equivalent circuit of the inductive v.t. is that given in Fig. 4.10. The inductor L_K in Fig. 4.12a is part of the PLC coupling circuit.

It can be seen from the simplified equivalent circuit of Fig. 4.12b that the lowest capacitor of the divider and the non-linear main inductance of the of the inductive v.t. IVT form a resonant circuit which can cause ferro-resonance. Such phenomena can be initiated, for example, by a fuse blowing in the secondary circuit of the v.t. The consequence is severe distortion of the secondary voltage due to sub-harmonic oscillations. Many circuits have been developed for preventing CVT's from oscillating; one of the simplest is to permanently connect a damping resistor across an auxiliary winding e-n. The total burden must not, of course, exceed the rated burden. A further method is to switch in loading capacitors using thyristors when resonance occurs, but this also has an influence on the transient behaviour of the CVT which can cause indiscriminate tripping in the case of

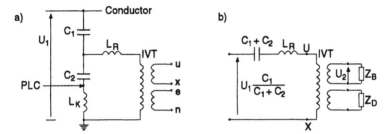

Fig. 4.12:   Basic circuit a) and simplified equivalent circuit b) of a CVT

fast protection devices. Should there be a step change of the primary voltage (e.g. as in the case of a close fault on the power system), the secondary voltage will contain an HF and/or a sub-harmonic oscillation depending on the instant of the fault (voltage zero or maximum), the capacitance of the divider and the CVT burden including the damping impedance Z_D [4.1, 4.3], which can adversely influence the operation of fast protection devices. While the HF oscillations can be filtered out of the measured voltage without great difficulty, the sub-harmonic oscillations in the range of 10 ... 25 Hz represent a particular danger. Features such as special filters can be included, which are capable of appreciable reducing sub-harmonic oscillations, but they can never be eliminated entirely.

As theoretical analyses and practical measurements have demonstrated, CVT's greatly reduce decaying DC components and HF oscillations.

4.2.3   Non-conventional v.t's

a) Capacitive voltage divider with amplifier

It is perhaps an obvious course of action to overcome the difficulties resulting from the parallel connection of the capacitor and inductor in a CVT by using an amplifier in place of the inductive v.t. to amplify the voltage tapped off from the capacitive voltage divider. A homogeneous voltage divider, the components of which all have the same phase-angle, will transform any waveform faithfully. The principle of this arrangement is shown in Fig. 4.13 and has performed well for both metering and protection purposes in iron-clad SF6 installations. The construction is simple and the transfer characteristics in the presence of transients good. The rated burden is generally lower than for conventional v.t's and for this reason, they are more frequently used in conjunction with modern protection devices having low burdens.

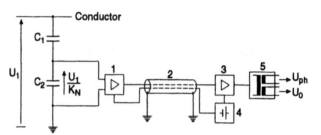

Fig. 4.13:   Basic circuit of a CVT comprising a capacitive voltage divider and solid-
state amplifiers [4.1]

The primary voltage is reduced to a low value U_1/K_N by the divider comprising C_1 and C_2 and applied to the pre-amplifier (1). The output voltage U_A of the pre-amplifier is conveyed by the coaxial cable (2) to the power amplifier (3) which is at ground potential. The signal at the output of this amplifier is connected to the primary of an isolating transformer (5). Under rated conditions, the voltages across the secondaries of the isolating transformer are the standard secondary values for phase and neutral voltages, i.e. $100/\sqrt{3}$ V and $100/3$ V.

The rated powers of this kind of CVT are typically a few tens of VA. The harmonics and HF oscillations contained in the primary voltage are transferred faithfully from a few Hz up to about 100 kHz [4.11]. For power system faults directly at the terminals, the secondary voltage falls to 1 % of rated voltage within 1 ms [4.1]. It is therefore particularly suitable for supplying protection devices with a fast response.

CVT's according to this principle have been in operation on HV systems for many years.

b) V.t's according to the Pockels effect

The linear electro-optical Pockels effect involves the measurement of the phase-shift δ between two light waves caused by the field strength E prevailing in a double refracting crystal. The phase-shift is given by

$$\delta_E = K_{eo} \cdot E \cdot l \tag{4.15}$$

where
K_{eo} = electro-optical coefficient; for quartz approximately $-8.12 \cdot 10^{-6}$ rad/Vm
E   = electrical field strength in the crystal in [V]
l   = length of the light path in the crystal in [m]

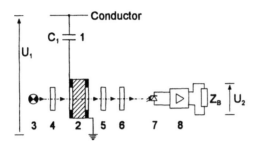

Fig. 4.14:   Basic circuit of a non-conventional v.t. based on the Pockels effect [4.1]

Figure 4.14 illustrates the principle of a v.t. based on the Pockels effect. A voltage divider comprising a capacitor C_1 and the Pockels cell of capacitance C_2 is connected across the voltage to be measured. The voltage across C_2 generates an electrical field proportional to the primary voltage. The beam of light from the source (3) is directed at the polarizer (4) which then produces two waves of light shifted in phase by $\pi/2$. These are directed at the Pockels cell (2). Because of the electrical field, the two waves of light have different speeds of propagation which gives rise to the phase-shift δ_E mentioned above. This phase-shift is shifted a further ¼ wavelength in the $\lambda/4$ plate (5) before being applied to the analyzer (6). The level of light at the output of the analyzer is proportional to the phase-shift δ_E and thus also to the primary voltage. The photo-diode converts the light intensity into a voltage which is amplified in (8) to produce the secondary voltage U_2.

Up to the present, only prototypes of this kind of v.t. have been tested and they have not be used in practice for supplying protection devices. The reason lies in the relative complexity of the design.

Part B
Analogue Protection

5

Solid-State Protection Relays and Systems

5.1 General

The first protection devices to follow fuses were electromechanical relays for detecting phase faults. As these relays developed, their application was extended to other kinds of faults [5.1]. The introduction of moving-coil measuring elements and rectifier bridges around 1950 [5.2, 5.3] accelerated operating times and reduced the burdens of the relays.

Solid-state relays — also referred to as electronic relays — appeared in about 1960 and presented the protection engineer with the following advantages:

- shorter operating and grading times
- greater accuracy
- better adaptation of the operating characteristic to suit power system conditions
- much reduced burden
- less maintenance
- longer life because of the absence of moving parts

Much effort, however, was needed to develop solid-state measuring and control circuits, which could stand up to the heavy-current environment and be able to operate right at the instant of a high energy fault as reliably as electromechanical relays.

64

In the meantime, the now usual modular design of solid-state equipment and contactless wiring techniques have enabled integrated protection systems to be constructed, which are mostly of the plug-in type.

Because of the prominence of solid-state relays today, this chapter on ana-logue protection only discusses solid-state techniques. Those electromechanical relays which are still in use are described in earlier publications [5.4, 5.5].

5.2 Basic design of solid-state protection devices

Solid-state protection devices are normally assembled from functional units which themselves are made up of standardized modules. This achieves [5.6]
- simplified testing using the user's test equipment
- easy fault-finding and correction using pre-tested replacement modules
- consistent standard interfaces
- simple adaptation to new requirements by adding modules

With reference to the basic block diagram of a solid-state protection device shown in Fig. 5.1, the input variables I (current) and U (voltage) enter the input module 1, which adjusts the signal levels for processing by the electronic circuits and provides the DC isolation between them and the primary system (c.t's and v.t's). The levels of the output signals from module 1 are then compared with ref-erence values in module 2, which picks up if they exceed the reference values and excites the timer 3. The logic 4 checks that 2 is still picked up when the time set for 3 expires and providing this condition is fulfilled excites the tripping block 5 and the signalling block 6. Module 7 provides the internal auxiliary supplies and module 8 is for testing the protection using either internal or external testing fa-cilities. This basic diagram varies according to protection task, protection crite-

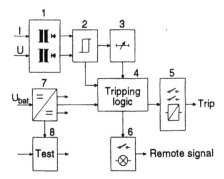

Fig. 5.1:   Block diagram of a solid-state protection device

ria, protected unit and series of modules [5.7, 5.8, 5.9]. Typical configurations are discussed below. The significance of the DIN symbols used is given in Appendix B.

5.3 Basic modules

5.3.1 Signal conditioning

5.3.1.1 Input transformers

The input transformers step down the secondary currents and voltages coming from the main primary system c.t's and v.t's to a level suitable for processing by the electronic circuits. At the same time they also provide DC isolation between the primary system c.t's and v.t's and the electronic circuits to prevent interference from reaching them. Figure 5.2a shows the simplified diagram of an input transformer for converting the impressed current coming from the main c.t. into a proportional DC voltage, Fig. 5.2b the corresponding three-phase arrangement and Fig. 5.2c the case of transformation to an AC signal without rectification.

5.3.1.2 Summation c.t's

Summation c.t's derive a single-phase signal from multi-phase inputs in order to economize on measuring circuits and to simplify the protection devices. They are

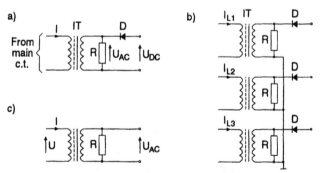

Fig. 5.2:   Basic single-phase circuit a) and three-phase version b) of input transformers with rectification of the input signal and a version without rectification c)

           IT   =   input transformer with static screen between primary and secondary

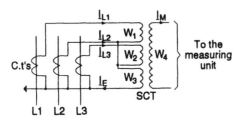

Fig. 5.3:   Circuit of a summation c.t. (SCT)

used mainly in connection with the longitudinal differential (pilot wire) protection of transmission lines and bus zone protection.

Figure 5.3 shows the basic circuit of a summation c.t. for obtaining a single-phase current $\underline{I}_M$ which is proportional to the three phase-to-neutral input currents. The single-phase signal is given by the geometrical sum of the input currents and the ratio of the summation c.t.

$$\underline{I}_M = k_1 \underline{I}_{L_1} + k_2 \underline{I}_{L_3} + k_3 \underline{I}_E \tag{5.1}$$

where $\underline{I}_{L_1}$ and $\underline{I}_{L_3}$ are the secondary currents of the c.t.'s in the phases L1 and L3 and $\underline{I}_E$ is the summation current, which is equal to zero for symmetrical load currents and faults not involving ground. In these cases (5.1) becomes

$$\underline{I}_M = k_1 \underline{I}_{L_1} + k_2 \underline{I}_{L_3} \tag{5.2}$$

In equations (5.1) and (5.2), the factors of proportionality k_1, k_2 and k_3 are determined by the turns ratio of the windings w_1, w_2 and w_3. Taking the turns of w_2 as reference, $k_1 = w_1/w_2$, $k = 1$ and $k_3 = w_3/w_2$. Typical choices in practice are $k_1 = 2$, $k_2 = 1$ and $k_3 = 3$. Equation (5.2) then becomes $\underline{I}_M = 2\underline{I}_{L_1} + \underline{I}_{L_3}$ and the rms value

$$I_M = \sqrt{3}\, I_{ph} \tag{5.3}$$

where I_{ph} is the rms value of the current in each of the phases.

It follows from equation (5.1) and Fig. 5.3 that the relative value of the current I_M in relation to the phase current I_{ph} varies depending on which phases are involved in a fault. The values of the ratio I_M/I_{ph} for the different kinds of faults in solidly or low-resistance grounded systems for $k_1 = 2$, $k_2 = 1$ and $k_3 = 3$ are listed in Table 5.1. It can be seen from the table that this arrangement is especially sensitive for ground faults, which represent the majority of faults in these power systems.

Table 5.1:   The value of the ratio I_M/I_{ph} for the different phase and ground faults
produced by the summation c.t. according to Fig. 5.3

Kind of fault	Phases involved	I_M/I_{ph}
Three-phase fault	L1 - L2 - L3	$\sqrt{3}$
Phase-to-phase faults	L1 - L2	2
	L2 - L3	1
	L3 - L1	1
Phase-to-phase faults	L1 - E	5
	L2 - E	3
	L3 - E	4

5.3.1.3 Symmetrical component filter

For certain kinds of protection it is necessary to extract the symmetrical components from the three-phase input signals. Figure 5.4 shows the basic arrangements of symmetrical component filters for current and voltage systems. In the
former case (Fig. 5.4a), the secondary currents of the main c.t's are converted in
the symmetrical component filter SCFI into a signal, which according to the
measurement task concerned is proportional to one or several of the symmetrical
components of the three phase currents. If, for example, all the components (i.e.
positive, negative and zero-sequence) are required for measurement, the output
signal I_M of the filter will be

$$I_M = k_1 I_1 + k_2 I_2 + k_0 I_0 \qquad (5.4)$$

where I_1, I_2 and I_3 are the positive, negative and zero-sequence currents (see
Section 3.6) and k_1, k_2 and k_3 complex coefficients of proportionality determined
by the choice of the internal circuit of the symmetrical component filter. Circuits
of this kind are used mainly in phase comparison protection schemes on HV
transmission lines to reduce the number of measuring systems and communication channels and also in schemes to detect asymmetric load (NPS protection).
Real numbers are chosen in the former case for the constants, i.e. $k_1 = 1$, $k_2 =$
3 ... 10 and $k_3 = 5$ [5.10, 5.11].

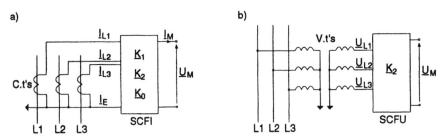

Fig. 5.4:   Combination of c.t's and v.t's with symmetrical component filters
a) symmetrical component filter for currents (SCFI)
b) symmetrical component filter for voltages (SCFU)

The zero-sequence component is eliminated in some phase comparison schemes and the output of the symmetrical component filter becomes $I_1 + k_2 I_2$ with $k_2 = 4 \ldots 10$.

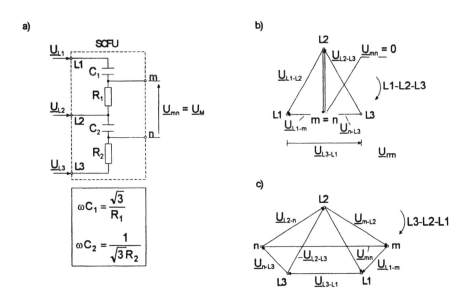

Fig. 5.5:   Circuit of a symmetrical component filter for voltage a) and corresponding vector diagrams for a symmetrical b) and asymmetrical c) three-phase system [5.12]

NPS protection uses either the negative-sequence component of the current or voltage to detect asymmetric load conditions and interrupted conductors. Figure 5.4b shows the connection of the v.t's to the symmetrical component filter and Fig. 5.5a the internal circuit of the filter. The corresponding vector diagrams for the normal load condition and during a fault are given in Figures 5.5b and 5.5c [5.12].

By appropriately selecting the values of R_1, R_2, C_1 and C_2, the output signal U_{mn} of the symmetrical component filter can be made to equal zero during normal symmetrical load conditions, while for an asymmetrical three-phase system $U_{mn} = \sqrt{3}\, U_N$, U_N being the secondary phase-to-phase voltage. An even simpler method is shown in Fig. 5.6 for an ungrounded synchronous generator. Only the phase currents $\underline{I}_{L1}$ and $\underline{I}_{L3}$ are connected to the symmetrical component filter, $\underline{I}_{L1}$ flowing through R_1 and $\underline{I}_{L3}$ through X and R_2 to generate the voltage drops $\underline{U}_{R1}$, $\underline{U}_X$ and $\underline{U}_{R2}$. The output voltage is thus

$$\underline{U}_M = \underline{U}_{mn} = \underline{U}_{R1} + \underline{U}_{XR}$$

$$\underline{U}_{XR} = \underline{U}_X + \underline{U}_{R2}$$

(5.6)

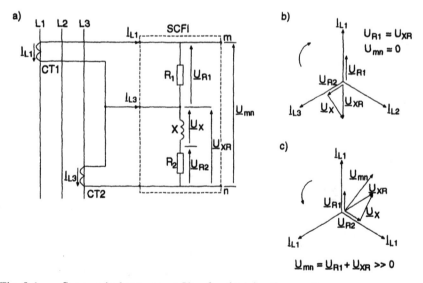

Fig. 5.6:   Symmetrical component filter for obtaining the negative-sequence component of the current in an ungrounded power system
a) circuit, b) vector diagram for the positive-sequence and c) vector diagram for the negative-sequence

The values of the resistors and impedances are chosen such that for a symmetrical load current $|U_{R1}| = |U_{R1}|$ and therefore $U_{mn} = 0$ (Fig. 5.6b). The output voltage U_{mn} for the negative-sequence system on the other hand is large (Fig. 5.6c).

5.3.1.4 Harmonic filter

Harmonic filters extract harmonics from the analogue input signals either to amplify and process them or to discard or attenuate them. For example in a biased differential scheme for a power transformer, the proportion of second harmonic in the current is monitored to determine whether the power transformer is being switched on or not and thus whether the protection needs to be restrained to prevent the inrush current from causing false tripping. In some instances, the fifth harmonic is also monitored in this kind of protection to guard against overfluxing due to overvoltages (e.g. after shedding load on power transformers at the end of long lines), which can similarly give rise to false tripping [5.13].

a) in Fig. 5.7 shows the frequency response curve of a bandpass filter used for this purpose and b) the circuit of the filter. The secondary winding of the input transformer IT and the capacitor C form a tuned circuit with resonant frequencies at 120 Hz and 300 Hz. The variable resistor is for setting the amplification ratio $S_{I/P}/S_{O/P}$.

The second kind of filter, which eliminates or attenuates all harmonics in the input signal but permits the fundamental to pass unchanged, is a low-pass filter.

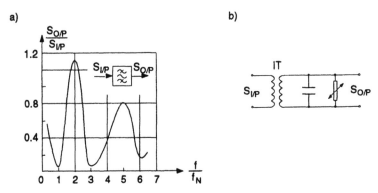

Fig. 5.7:   Frequency response curve a) and circuit b) of a bandpass filter for the second and fifth harmonics

$S_{I/P}$ = input signal $S_{O/P}$ = output signal

a)   b)

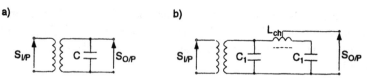

Fig. 5.8:   Filter circuits for suppressing all harmonics a) or a particular harmonic b)

A simple low-pass filter is shown in Fig. 5.8a. The tuned circuit in this case is once again formed by the inductance of the secondary winding of the input transformer and the capacitor, but the resonant frequency this time is the power system frequency.

The passive bandpass filter shown in Fig. 5.8b extracts a particular harmonic from the input signal.

The circuits of other analogue filters and the derivation of their characteristics are to be found in the literature, e.g. [5.14]. Digital filters are described in Chapter 11.

5.3.2   Measuring principles

5.3.2.1 Measurement of a single variable

Circuits which measure a single variable monitor a limit by comparing the amplitude of the input variable with a defined, usually adjustable pick-up setting.

In solid-state protection devices, the limit monitors are mostly Schmitt triggers, which have an analogue input signal and a binary output signal. The operating principle is illustrated in Fig. 5.9. The output signal $S_{O/P}$ equals zero, respectively the signal level defined as "logical 0", for as long as the instantaneous value of the input signal $S_{I/P}$ remains below the pick-up setting S_{pu} (Fig. 5.9a and b). Whenever $S_{I/P}$ exceeds the setting S_{pu}, the voltage at the output of the trigger switches to the potential defined as "logical 1" and remains at that potential until $S_{I/P}$ falls below the reset value S_r, at which point the output signal $S_{O/P}$ reverts to "logical 0". The output signal is thus a squarewave, the duration of which is determined by the time the input signal is higher than the pick-up setting and the reset value. If as with electromechanical relays a reset ratio is defined as

$$RR = \frac{S_r}{S_{pu}} \qquad (5.7)$$

where S_r is the reset value and S_{pu} the pick-up setting, then the reset ratio RR of overcurrent and overvoltage relays must be less than 1. In contrast to electromechanical relays, the reset ratio of solid-state relays can be chosen very close to 1,

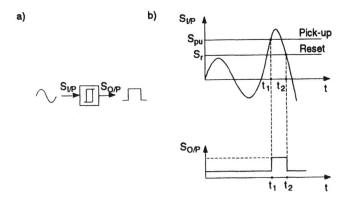

Fig. 5.9:   Limit monitor for a single input variable
              a) basic circuit   b) operation when the input signal exceeds the setting

but it must be borne in mind that the output signal oscillates for reset ratios higher than 1 and the trigger can no longer perform its function.

It follows from this explanation of the operation that a trigger monitors the instantaneous value of the input variable, while electromechanical relays always integrate the input signal in some form or other and respond to the integrated value.

By simply inverting the output signal (reversal of the signal or of logical 1 and 0), the output can be made to remain at logical 1 for as long as the input signal is below the reset value and does not reach the pick-up value. Thus the distinction which had to be made between electromechanical under and overvoltage relays with regard to the continuous rating of the undervoltage relay does not apply to solid-state relays, because the only difference between under and overvoltage functions is the definition of which logical signal shall be processed. The trigger function can also be performed by an operational amplifier according to Fig. 5.10.

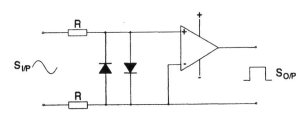

Fig. 5.10:   Limit monitor function performed by an operational amplifier working as a shaper with squarewave output [5.14]

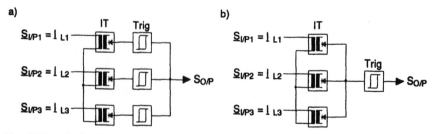

Fig. 5.11:   Individual a) and common measurement b) of the three phase currents

With the aid of appropriate input networks, limit monitors can also monitor several input variables. Summation c.t's were discussed in Chapter 4, but rectifiers also enable, for example, the greater of the three phase currents to be detected using a single limit monitor (maximum value detector). Thus with maximum value detection, what is basically a single-phase overcurrent relay according to Fig. 5.11b can serve the same purpose as the three-phase overcurrent relay according to Fig. 5.11a.

5.3.2.2 Measurement of a several variables: amplitude comparators

Measuring elements for two variables monitor either their product or their quotient. The two input variables are mostly current and voltage or two currents, or combinations thereof. There are basically two kinds of comparators, amplitude comparators and phase comparators.

An amplitude comparator is shown in Fig. 5.12. The two input variables $S_{I/P1}$ and $S_{I/P2}$ may be two currents, two voltages or a current and a voltage. The essential thing is that one of the two has an operating influence on the measuring system (limit monitor) and the other a restraining influence (i.e. prevents tripping). Therefore the limit monitor only produces an output signal $S_{O/P}$ when

$$|\underline{S}_{I/P1}| > |\underline{S}_{I/P2}| \qquad (5.8)$$

where
$S_{I/P1}$ = operating input variable
$S_{I/P2}$ = restraining input variable

Amplitude comparators for two input variables are used in differential and impedance relays.

Since to conform to equation (5.8) two absolute values have to be compared, the input variables must be rectified and smoothed. The same task can be performed on the other hand by phase comparison, providing the phase relationship between the sum and the difference of the two variables is evaluated.

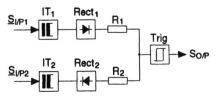

Fig. 5.12:   Principle of an amplitude comparator for two input variables

5.3.2.3 Measurement of a several variables: phase comparators

Phase comparators monitor the phase-angle between the input variables and generate an output when it exceeds or falls below a set angle. A simplified diagram is shown in Fig. 5.13. The input variables $S_{I/P1}$ and $S_{I/P2}$ are first shaped into squarewave impulses $S'_{I/P1}$ and $S'_{I/P2}$ and then the time t_k between them is monitored. In relation to a fundamental with an angular velocity ω_N where $\omega_N = 2\pi f_N$, this time is proportional to the phase-angle α_k between $S_{I/P1}$ and $S_{I/P2}$ since $t_k = \alpha_k/\omega_N$. This is also referred to as a coincidence measurement. If a limit time t_G is set on the timer T, an output signal is generated when the coincidence $t_k > t_G$, i.e. when the phase-angle $\alpha_k = \alpha_G$ where $\alpha_G = t_k$. Assuming $\alpha_G = \pi/2$, $t_G = 4.16$ ms (at 60 Hz) and an output signal is generated for $t_k > 4.16$ ms.

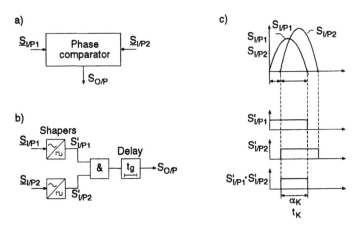

Fig. 5.13:   Simplified block diagram of a phase comparator a) and b) and the comparison of the signals c)

Figure 5.13c shows the comparison of the signals for the positive half-cycles. The same arrangement can be used for the additional comparison of the negative half-cycles, which makes measurement faster and more reliable. Harmonics, DC component etc., must be removed from the input variables before the can be used for measurement.

A alternative method [5.15] is given in Fig. 5.14, in which the output signal is dependent on the comparison of the coincidence time t_k with the anti-coincidence time t_{ak} (i.e. the time during which the two squarewave signals are of opposite polarity).

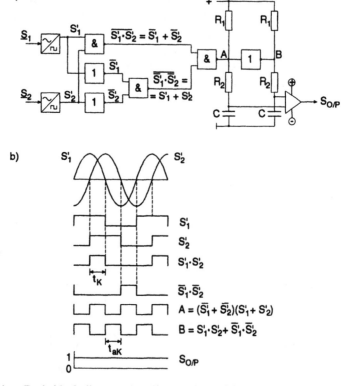

Fig. 5.14:   Basic block diagram a) and comparison of the signals b) in the case of a combined coincidence and anti-coincidence measurement

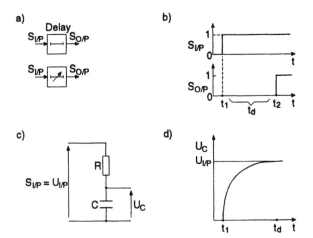

Fig. 5.15:   Simple solid-state timer:
a) symbols used for a timer with a fixed setting and one with a variable
setting;   b) timing diagram of the signals; c) R/C circuit; d) excursion
of the voltage U_C

5.3.3   Timers

Timers issue an output signal $S_{O/P}$ a set time t_d after the input signal $S_{I/P}$ is applied to the input. This function is usually achieved in analogue solid-state circuits using an R/C arrangement according to Fig. 5.15. If the input signal $S_{I/P}$ is a DC voltage $U_{I/P}$, the output signal $S_{O/P} = U_{O/P}$ where

$$U_{O/P} = U_{I/P}\left(1 - e^{-\frac{t}{RC}}\right) \tag{5.9}$$

The time constant of the charging circuit and therefore the time delay t_d are determined by the values of R and C.

    Timers with fixed settings are used, for example, in phase comparators according to Fig. 5.13 and timers with variable settings in all time-graded protection schemes.

5.3.4 Auxiliary supply units

The modules of solid-state protection devices require a number of auxiliary supplies, which are usually provided by power supply units installed in the protection devices themselves. The power supply units draw their power either from the same station battery used to supply the circuit-breaker tripping coils actuated by the protection, or from a battery which is at least as reliable. The power supply units used in protection are normally DC/DC converters in order to DC isolate the electronic circuits from the station battery and the switching transients to which it is subject every time a circuit-breaker is operated. Generally the power supply units also include some form of energy reservoir to maintain the supply to the protection during voltage dips on the battery supply (some tens of ms). The output voltages provided by the power supply units usually have narrower tolerances than the battery voltage, so that the electronic circuits can be designed for a lower power dissipation and therefore to be smaller than would otherwise be the case.

Figure 5.16 shows a simplified diagram of a power supply unit as used for protection, comprising a DC energy reservoir (Res) and a DC/DC converter. The reservoir Res includes filter F to block interference superimposed on the battery voltage U_B. The capacitor C bridges short interruptions of the battery supply and must be large enough to maintain the supply to the protection for the time taken by fuses etc. to clear short-circuits on the other loads connected to the battery. C is charge via the resistor R and discharges via the diode D.

The input voltage $U_{I/P}$ is chopped in the DC/DC converter into squarewave voltage with a frequency of a few kHz and then stepped down by the transformer T, which also provides the DC isolation between input and output. The secondary voltage is then rectified and smoothed [5.16, 5.17].

Figure 5.17 shows a power supply unit of similar design, but with the functions divided into separate modules, which in different combinations enable a whole series of power supply units to be put together for different requirements [5.18, 5.19].

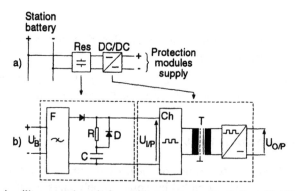

Fig. 5.16:   Auxiliary supply unit for solid-state protection devices [5.17]
a) block diagram   b) simplified circuit

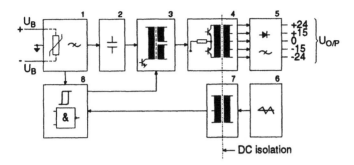

Fig. 5.17:   Block diagram of an auxiliary supply unit [5.18]

The battery supply first passes the RF filter and overvoltage protection (1) before charging the energy reservoir (2). The input voltage is then converted into an AC by the transistors and transformer of the switching regulator (3), which generates an intermediate voltage of 200 V and two stabilized auxiliary voltages. The switching regulator is controlled by the supervision and control unit (8) which also includes the start-up circuit. The supervision and control unit enables the power supply unit to start-up after an interruption of the battery supply and also limits the output current in the event of a short-circuit in the electronic modules. The reservoir is a robust electrolytic capacitor which also reduces any ripple on the battery supply. The outputs of the push-pull stage (4) are the voltages needed for the electronic circuits. The pulse width regulator (6) monitors the voltages of the rectifier and smoothing circuit (5) and transmits control impulses via the impulse transformer (7) to the supervision and control unit (8), thus forming a fast, accurate and almost load insensitive regulation loop for the output voltages. The push-pull transformer in (4) and the impulse transformer in (7) provide the DC isolation between battery and electronic supplies. The availability of all the output voltages is continuously monitored.

5.3.5   Tripping logics

In the case of large protection systems with a number of protection functions, the outputs of the various functions are grouped and assigned to the different tripping signals by a tripping logic. Typical examples are the combination of starting and measuring unit signals in distance relays and the tripping and signalling channels of a generator protection for a generator/transformer set, where the output channels are energised by different combinations of several protection functions. The output channels can be tripping commands, alarms, signals etc.

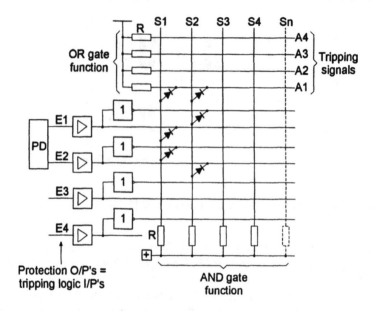

Fig. 5.18:   Simplified diagram of a programmable tripping logic [5.21]

The tripping logics of the simpler protection devices are generally hard-wired, whereas those of larger systems can be programmed, e.g. by means of a matrix with diode plugs. A simplified representation of the latter can be seen in Fig. 5.18. PD is the protection device with output signals of positive polarity. The upper part of the logic forms an OR gate and the lower part an AND gate. In the example shown, the tripping channel A1 will only be positive, if one and only one of the two inputs E1 and E2 is positive, i.e. A1 = 1 when E1 = 1 and E2 = 0 or when E1 = 0 and E2 = 1, which can be written $A1 = \left(E1.\overline{E2}\right) \vee \left(\overline{E1}.E2\right)$. It can be seen that the first bracket is formed by the rail S1, which can only be positive if E1 is positive and E2 is zero etc.

In practice, a programmable tripping logic can have several tens of inputs and up to twenty outputs [5.20, 5.21], e.g. the protection systems for generator/transformer sets.

5.3.6   Tripping units

Tripping units convert the low-power output signals of the measuring and logic units into several heavy-duty DC isolated tripping signals. Surge circuits or the application of extremely high-speed auxiliary tripping relays enable tripping units

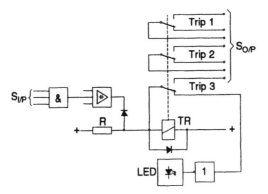

Fig. 5.19:   Simplified diagram of a tripping channel in a tripping unit for increasing the power and number of tripping signals
TR = auxiliary tripping relay   LED = light emitting diode

to achieve operating times of a few milliseconds. Thyristor tripping units are in use in a number of countries. Figure 5.19 shows an example of a tripping channel which increases the power and number of signals between input and output. In practice, a tripping unit contains several of these circuits which can have different input logics.

5.3.7   Signalling units

Signalling units are in principle the same as tripping units with the exception that the ratings of the signalling contacts are normally lower than those in tripping units, they also provide, however, DC isolation between the measuring circuits and the outputs. They comprise as a rule - possibly with pre-amplifiers - auxiliary relays either of the attracted armature (Fig. 5.20a) or reed type (Fig. 5.20b). The operating times are of the order of just a few milliseconds.

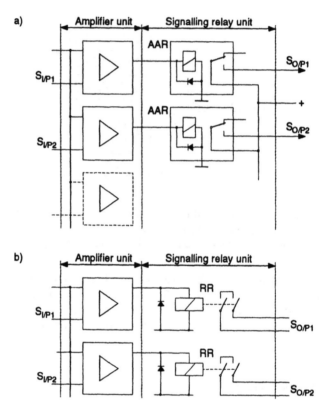

Fig. 5.20:   Examples of signalling units comprising attracted armature auxiliary relays a) and reed type auxiliary relays

5.3.8   Opto-coupler units

Opto-couplers are used to provide DC isolation between incoming binary signals and the internal circuits of the protection. The principle can be seen from Fig. 5.21, in which output 2 is also equipped with an inverter.

An opto-coupler comprises a light emitting diode and a photo-transistor, the light generated by the LED impinging on the photo-transistor. The LED lights when a DC voltage signal is applied to the corresponding input. The light causes the photo-transistor to conduct a current between emitter and collector, which is amplified and converted to a suitable voltage signal for processing by the internal circuits of the protection [5.22]. The light of an opto-coupler provides DC isolation as does an auxiliary relay, but is much faster.

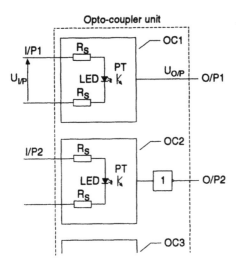

Fig. 5.21:   Principle of an opto-coupler [5.22]
             OC = opto-coupler PT = photo-transistor

5.3.9   Units for continuous monitoring and testing

Solid-state relays have the advantage of being able to continuously monitor the status of the most important circuits and modules and thus always be able to provide information as to the availability of the protection. The simplest monitor checks that the auxiliary supply voltages are not out of tolerance and blocks the tripping circuits to prevent false tripping and gives alarm if one of them is. The principle is illustrated in Fig. 5.22. The internal auxiliary supplies U_1 and U_2 (e.g. +24 V and +15 V) are each monitored by two limit detectors. If one of the supplies moves in one direction or the other out of the permissible tolerance band, an auxiliary relay (Alarm) is energised via an amplifier after a delay of a few milliseconds, the contacts of which interrupt all the tripping circuits and give alarm.

In larger protection systems, external and internal comparators, logics, the integrity of the tripping circuits, the symmetry of input variables etc., are also monitored [1.2, 5.23].

Similar continuous monitoring of electromechanical relays was only possible to a much lesser extent and therefore periodically testing them was essential. Although this is no longer necessary with solid-state relays, there still remain some circuits that cannot be covered by continuous monitoring and therefore have to be tested.

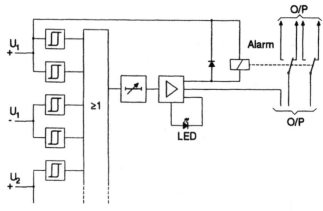

Fig. 5.22:   Continuous monitoring of the auxiliary supplies in solid-state protection devices [5.23]

Two kinds of test procedures are common, one which qualitatively checks that a circuit is able to function and another which quantitatively checks the pick-up level. In contrast to electromechanical relays, both these tests can be performed during normal operation in the case of solid-state relays. It is even possible to repeatedly check the pick-up level automatically. Although automatic testing takes very little time, the protection is briefly out of service while testing is proceeding and therefore it is of advantage for the protection functions to be divided into two basically equivalent groups which are then tested alternately. This is the technique, for example, in the case of large generators, where one group is taken out of service for testing while the other group continues to protect the generator, and then the second group is tested after the first is back in service and so on [5.25].

5.4 Combination of the modules and units to protection devices

Solid-state protection devices or their modules can be relatively simply combined to form larger protection devices to perform the various protection tasks. These are described in the next chapter. The examples given below are intended to explain the general principles.

Figure 5.23 shows possible configurations of standard modules and functional units to achieve an overcurrent protection relay a) and a distance protection relay b). To what extent the modules and functional units actually exist as hardware or

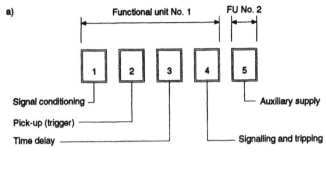

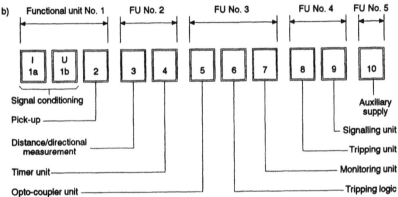

Fig. 5.23:   Combination of solid-state modules to form functional units and complete
protection devices
a) overcurrent protection   b) distance protection

are standard circuits in a software library varies from one application to another,
from one manufacturer to another, how often the function is needed etc.

5.5   Combination of protection devices to protection systems

The logical continuation of combining functions to protection devices is to combine protection devices for a particular item of primary plant to a protection system. It depends again on the factors given above and also on the requirements
with respect to degree of redundancy and back-up protection, whether the final
solution is a system with several measuring units or a single measuring unit with

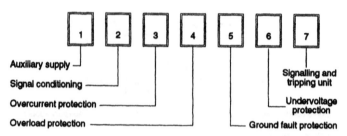

Fig. 5.24:   Configuration of an integrated protection system for an HV motor
              comprising several protection devices

a variety of input circuits. For example, assuming an HV induction motor according to [5.24] has to be protected against the following kinds of faults

- phase faults
- thermal overload (during starting and operation under load)
- stator ground faults
- single-phasing (interrupted conductor)

at least the following protection devices are necessary

- overcurrent
- thermal overload
- ground fault
- overvoltage

Each one of the above protection devices can be assembled according to Section
5.4 (Fig. 5.23) and supplied by a common auxiliary supply unit and actuate a
common tripping unit (Fig. 5.24). It is also possible - and sensible in the case of
large quantities of the same system - to fully integrate the individual protection
devices to form a single device, in which case the same input variables are used
for several measurements, measuring elements can be combined etc.

    Not always to be recommended, but sometimes justifiable for reasons of cost
is the combination of protection devices for different items of primary plant as is
shown in Fig. 5.25 for a group of HV auxiliary drives in a power plant. The obvious operational disadvantage of such an arrangement is that when one item of
plant is out of service (e.g. for maintenance), its protection is still in operation
and capable of generating tripping signals.

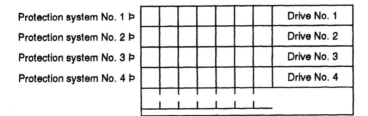

Fig. 5.25:   Combination of the protection devices for several HV drives

6

Application of Analogue Protection Devices

The most important criteria used for detecting faults were described in Chapter 3 without going into the design of the protection devices themselves. This was deliberately omitted, because depending on the type of measuring system employed, i.e. analogue or digital, their operating principles are quite different and therefore have to be dealt with separately. This chapter is devoted to the most common analogue protection devices. The more specialized devices which measure less frequently used system variables or are only applied for specific items of primary plant are discussed in Chapters 7 and 8.

6.1 Overcurrent protection

A distinction is made between instantaneously operating (i.e. without intentional delay) and delayed operating overcurrent protection (time-overcurrent relay). In certain instances, a time-overcurrent device is augmented by a directional unit or an undervoltage device. All the overcurrent schemes described below are for the detection of phase faults or ground faults.

6.1.1 Instantaneous overcurrent protection

Secondary relays, which measure the current flowing through the protected unit at a reduced level via current transformers, were introduced as an alternative to HV fuses and electromechanical trips relative early in the history of protection engineering. Their pick-up current must be set safely above the maximum load current to be expected on the one hand and below the minimum fault current on the other. A typical electromechanical overcurrent relay had a minimum operating time in the region of two to five periods of the power system frequency (33 ms to 83 ms) and in spite of the desire to make it faster, the pick-up setting often had to be set relatively high to avoid false tripping during switching operations on the primary system (i.e. due to the inrush currents of power transformers, capacitor banks or squirrel-cage motors) and the less sensitive setting tended to make the relay slower. This was one of the reasons why in many cases overcurrent relays were subsequently replaced by more selective kinds of protection such as differential or distance protection.

With the advent of solid-state protection devices, the minimum operating time of overcurrent relays was reduced to just a few milliseconds, which meant that the pick-up setting for certain items of primary plant had to be raised still further with the result that the protection became even less sensitive. Thus the application of solid-state instantaneous overcurrent relays is mainly restricted to the phase fault protection of power transformers with low ratings (less than 5 MVA) and HV induction motors. In the latter case, however, a short time delay of 50 ms is frequently necessary to prevent false tripping during inrush currents [6.1].

Figure 6.1 shows the schematic of a two-phase solid-state overcurrent protection for a protected unit of the kind mentioned above. The overcurrent relay Rel monitors the current of the protected unit my means of the c.t's. The fault current of a phase-to-phase or three-phase fault downstream of the relay exceeds its pick-up setting and the relay energises its tripping O/P. The resulting tripping signal S_{trip} excites the circuit-breaker tripping coil TC via an auxiliary contact on the circuit-breaker. A two-phase arrangement is sufficient to detect all phase

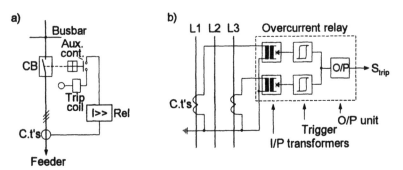

Fig. 6.1:   Schematic a) and block diagram b) for two-phase instantaneous overcurrent protection

faults in ungrounded or high resistance grounded systems or in systems grounded via Petersen coils. Low resistance or solidly grounded power systems require a full three-phase scheme.

The reach of instantaneous overcurrent protection devices depends on the pick-up current setting and the magnitude of the fault current. This is illustrated for the example of a distribution transformer (Fig. 6.2) which is supplied at a constant short-circuit power S_F. To maintain proper discrimination (i.e. detect faults selectively), the pick-up setting I_{set} of the protection must be high enough to prevent the protection from operating for a fault in or beyond station B. This is only the case for

$$I_{set} > I''_{f3pB} \tag{6.1}$$

where I''_{f3pB} is the initial three-phase fault current for faults in station B. This limits the protected zone, however, to the relative reactance of the transformer, i.e. $\alpha_3 X_T$ for a three-phase fault and $\alpha_2 X_T$ for a phase-to-phase fault, where α is the relative number of turns of the transformer. Since such small part of the transformer winding is protected, the disadvantage of this kind of protection is obvi-

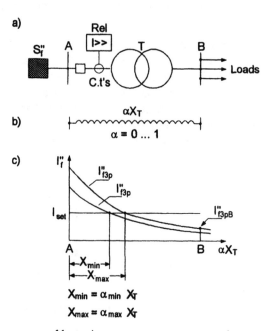

Fig. 6.2:   Zone protected by an instantaneous overcurrent relay
a) primary system;   b) relative transformer reactance;   c) fault current and protection reach

ous. The situation is much the same for other kinds of protected unit. The part of the power transformer not protected by an instantaneous overcurrent relay can be protected by delayed overcurrent relay which can have a much lower pick-up setting.

6.1.2   Time-overcurrent protection

Time-overcurrent protection is mainly applied to achieving discrimination, i.e. selective operation, for faults on lines and in transformers in radial MV power systems. But where discrimination is not the main objective, a short delay does permit a lower pick-up setting and therefore higher sensitivity if the application requires it.

There are two kinds of time-overcurrent relays, definite time and inverse time. In the case of the former, the operating time is constant regardless of the current flowing through the relay providing it is higher than the pick-up setting. The operating time of an inverse time-overcurrent relay, on the other hand, is inversely proportional to the level of the fault current. Fig. 6.3 shows the operating characteristics of both types and Fig. 6.4 how they would be time-graded for discriminate detection of faults in a radial power system (Fig. 6.4a). The operating times for faults at different locations using definite time-overcurrent relays can be seen from Fig. 6.4b and using inverse time-overcurrent relays from Fig. 6.4c. Discrimination in the first case is a matter of grading (staggering) the time delay settings, while in the second case the time/current characteristics of the relays have to be coordinated.

Considering firstly the definite time-overcurrent scheme, the calculation of the time delay to be set on each relay starts with the one at the remote end of the radial line, i.e. at the load. As can be seen from Fig. 6.4b, the time setting is increased by Δt for each relay working backwards towards the source. The grading time Δt is generally between 0.3 s and 0.5 s and represents the sum the fault

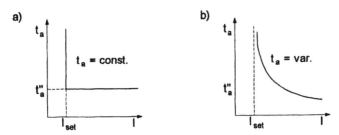

Fig. 6.3:   Operating time in relation to current for a definite time-overcurrent relay a) and an inverse time-overcurrent relay b)

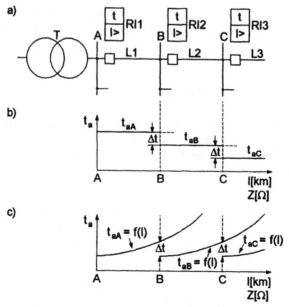

Fig. 6.4:   Discriminate protection of a line in a radial power system a) by definite
             time-overcurrent relays b) and by inverse time-overcurrent relays c)

clearing time of the next downstream relay (relay operating time + circuit-breaker
tripping time), the time taken by the downstream relay to reset after the fault current has been interrupted and a safety margin to cover the relay and circuit-
breaker tolerances. For the case of relay RI2 at the beginning of the section of
line L2, the corresponding time delay $t_{aB} = t_{acmax} + \Delta t$, where t_{acmax} is the longest
delay set on a time-overcurrent relay in station C.

The disadvantage of the time grading principle is that faults near the source
which have a higher fault level than faults on sections further down the line are
tripped after the longest delay. The destructive potential of a fault close to the
source is thus several times greater. As can be seen from Fig. 6.4c, the second
kind of time-overcurrent protection using relays with an inverse characteristic,
otherwise referred to as IDMT (inverse definite minimum time) relays, does not
suffer from this drawback, because their time delay reduces as the fault current
increases. The practice in this respect is somewhat divided, some countries
preferring definite time-overcurrent relays to avoid the more complicated grading
of the inverse characteristics of the various relays in a power system, while others
have standardized on IDMT relays [6.2]. Today, the grading of IDMT relays is
frequently performed by computer programs.

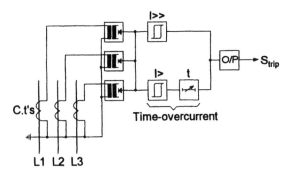

Fig. 6.5:   Three-phase time-overcurrent relay with instantaneous high-set stage

The pick-up current of definite time-overcurrent relays is chosen such that it is higher than the maximum load current I_{Lmax} of the protected unit and lower than the minimum initial AC fault current at the relay location for a fault at the end of the main zone of protection. For the case of the line L1 in Fig. 6.4, this means that

$$I''_{FminB} > I_{set} > I_{Lmax} \qquad (6.2)$$

where I''_{FminB} applies for a phase-to-phase fault at the end of the line L1. This setting, of course, is still a primary value and has to be converted to a secondary setting by dividing I_{set} by the nominal ratio K_N of the c.t's.

In certain cases, the protection of a radial system can be improved by a combination of instantaneous overcurrent relays as described in the preceding Section and time-overcurrent relays. The block diagram of a time-overcurrent relay with an instantaneous high-set stage is shown in Fig. 6.5.

In a scheme of this kind, which is used mainly on radial feeders and for smaller power transformers in MV systems, the largest current of the phase or phases involved in a fault (maximum value detection) is applied to the triggers of time delayed and instantaneous stages.

Finding satisfactory settings for a mixture of definite time and inverse time-overcurrent relays in the same power system is extremely difficult or even impossible.

6.1.3   Voltage controlled time-overcurrent protection

In certain instances, the overcurrent criterion does not always exist for a sufficiently long time to conclusively detect faults and a voltage criterion is added. The combination is referred to as voltage controlled time-overcurrent protection. Typical applications are low power synchronous generators (less than 100 MW), power transformers operating in parallel, railway supply systems etc., subject to brief overloads. A brief overload on a synchronous generator can be caused, for example, by automatic overexcitation in response to a fault in another part of the power system and in the case of power transformers in parallel, the one remaining in operation is clearly overloaded if one of the pair is tripped. In none of these instances may the overcurrent protection, which is only intended to detect fault conditions and not overloads, be allowed to trip. The stability of the overcurrent protection is achieved by adding the undervoltage criterion to distinguish between an overload and a fault condition, the former causing no appreciable voltage reduction and the latter always being accompanied by a significant voltage drop. A voltage controlled time-overcurrent protection is thus a combination of protection devices which can only trip the protected unit when both the overcurrent and undervoltage criteria are fulfilled at the same time. The same purpose is achieved by an underimpedance relay which may be considered as an overcurrent function with continuously variable pick-up setting controlled by the voltage. A simplified diagram of a voltage controlled time-overcurrent scheme for a power transformer and a synchronous generator is shown in Fig. 6.6. The scheme operates discriminately and is generally a three-phase protection, i.e. it measures the currents and voltages of the three phases.

The setting of the undervoltage unit must satisfy the following condition

$$\frac{U_{Lmin}}{K_N} > U_{set} > \frac{U_{Fmax}}{K_N} \tag{6.3}$$

where

U_{set} = pick-up setting of the voltage unit

U_{Lmin} = lowest permissible voltage at the relay location under load conditions

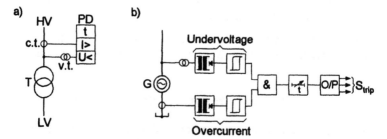

Fig. 6.6:   Voltage controlled time-overcurrent protection for a power transformer a) and a synchronous generator b)
HV = high-voltage system   LV = low-voltage system

U_{Fmax} = maximum voltage at the relay location for a fault at the end of the zone of protection

K_N = nominal v.t. ratio

The pick-up setting of the overcurrent unit is determined according to condition (6.2) and the time delay according to the time grading procedure of Section 6.1.2.

6.1.4   Directional time-overcurrent protection

As intimated in Section 3.5, the overcurrent criterion requires the additional information of the direction in which fault energy is flowing in order to operate discriminately in ring or meshed power systems. This is also necessary for double circuit lines and parallel power transformers in what are basically radial systems. It was demonstrated for a double circuit line (Fig. 3.3a) that directional overcurrent devices must be used at the relay locations 2 and 4 in station B. The situation would be much the same, if stations A and B were linked two power transformers instead of the lines L1 and L2.

Before describing the operating principle of a directional time-overcurrent scheme which is basically the same for items of plant in parallel and for ring lines, an explanation is given of the procedure for determining the settings for the directional time-overcurrent protection of double circuit lines and power transformers in parallel.

In order to prevent the protection of the healthy line or power transformer from tripping as the result of the overload which occurs immediately after the parallel unit is tripped, the pick-up current I_{set} must be chosen to satisfy the condition (6.2), whereby the maximum load current I_{Lmax} in this case is the sum of the loads currents of both lines or both transformers, i.e.

$$I_{Lmax} = I_{LmaxL1} + I_{LmaxL2} \quad \text{respectively} \quad I_{Lmax} = I_{NT1} + I_{NT2}, \, I_{NT1}$$

where I_{NT1} and I_{NT2} are the rated currents of power transformers T1 and T2. The sensitivity of the protection is adequate providing the current flowing through the protection for a fault at the end of the protected line, respectively the downstream side of the transformer is at least 50 % higher than the pick-up current. Considering protection PD2 in Fig. 6.7, for example, this means that the minimum initial fault current for a phase-to-phase fault at the remote end of the line with CB1 open (fault location F_A) is taken as reference value.

The time delays t_{a2} and t_{a4} set on the directional time-overcurrent units PD2 and PD4 in such a case would be a few tenths of a second (usually 0.3 to 0.5 s). Theoretically no delay is necessary at all at these locations, because PD2 and

PD4 can only trip for a fault on the line they are protecting, i.e. when the flow of fault energy is from station B to station A. The directional units interlock the tripping signal of the corresponding protection devices when the fault energy flows from station A to station B. A short time delay is included for safety reasons to exclude any risk of indiscriminate operation as a consequence of electromagnetic balancing phenomena during faults.

As can be seen from Fig. 6.7b, the time delays of the non-directional time overcurrent devices PD1, PD3 and PD5 are determined according to the normal time grading procedure.

Directional time-overcurrent devices must also be installed at certain points on a ring line to maintain discrimination for all faults. It must be emphasised, however, that this kind of protection is only suitable for ring lines with a single infeed. At which points directional and non-directional units are installed depends on the configuration of the power system and the time delays of the individual devices. The procedure is explained with reference to the simple ring system of Fig. 6.8a.

a)

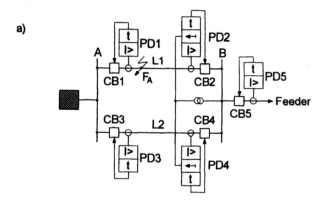

b)

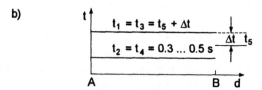

Fig. 6.7:   Directional time-overcurrent protection of a double circuit line a) and the corresponding grading diagram b)

t1...5   = time delays of the protection devices PD1 to PD5

d   = distance

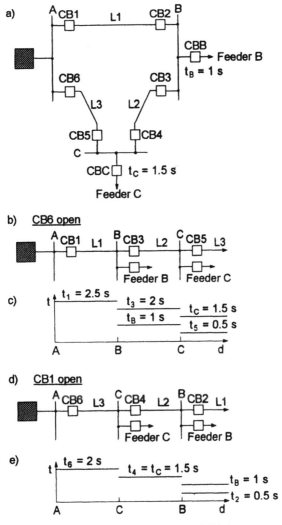

Fig. 6.8:   Time grading time-overcurrent relays in an MV ring system

For purposes of calculation, the ring line is initially interrupted at the infeed station A to make a radial line, e.g. by opening the circuit-breaker CB6. Only the circuit-breakers at the beginning of the sections of line are considered at this stage as illustrated in Fig. 6.8b. Assuming that lines L1 to L3 and the outgoing feeders in the substations are only equipped with time-overcurrent units, they can be time graded as described in Section 6.1.2. It is also assumed that the time set on the protection at location 5 is 0.5 s. Taking due account of the outgoing feeders in stations B and C and applying a grading time Δt of 0.5 s, the grading diagram of Fig. 6.8c is obtained. The timer settings for the remaining overcurrent devices are achieved in the same way by closing circuit-breaker CB6 and opening CB1. In this case, the system configuration becomes that of Fig. 6.8d and the grading diagram Fig. 6.8e.

The protection devices for the different locations now have to be chosen. Starting with the infeed station where all the feeders are always equipped with non-directional time-overcurrent units, the example under consideration requires that definite time-overcurrent relays with timer settings of $t_{a1} = 2.5$ s and $t_{a6} = 2$ s be installed at locations 1 and 6. Those stations and locations where directional units are needed are determined by comparison of the time delays of the various feeders. They are installed to preserve discrimination where the time delays are shorter than the delays according to the grading diagrams (Figures 6.8c and

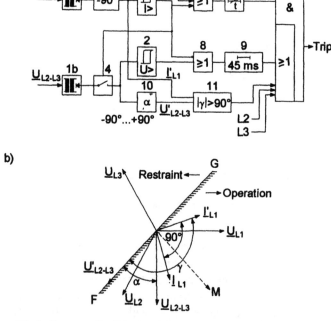

Fig. 6.9:   Block diagram of a directional time-overcurrent scheme a) and corresponding vector diagram b) [6.3]

6.8e). Considering the two relays in station B, this rule requires that a directional unit must be added for line L1, since with a setting of 0.5 s, t_{a2} is shorter than the 2 s of t_{a3}. Similarly, in station C, $t_{a4} = 1.5$ s and $t_{a5} = 0.5$ s so that just line L3 needs a directional unit. Because, however, the same times were determined for line L2 and the outgoing feeder C, i.e. $t_{a4} = t_{ac} = 1.5$ s, L2 in station C must also be equipped with a directional unit to prevent false tripping of circuit-breaker CB4 for a fault on feeder C. Thus of six locations where time-overcurrent relays are installed, three have to be directional (locations 2, 4 and 5) and three non-directional (locations 1, 3 and 6) and the timer settings are as given in the grading diagrams. The pick-up currents of all the relays are determined according to the relationship (6.2).

The operating principle of a directional time-overcurrent scheme is explained with reference to the block diagram and vector diagram of Fig. 6.9 [6.3]. The block diagram only shows the functions needed for one phase (L1). A complete scheme has three separate input and measuring circuits and only items 6, 7 and 12 are common for the three phases.

From Fig. 6.9a it can be seen that the amplitude of the phase current I_{L1} is first adjusted in amplitude by the input transformer (1a) and rotated by -90° in the phase-shifter (2) before being applied to the overcurrent detector (3) and the phase comparator (11) (directional measurement). The second input variable needed by the phase comparator to determine power direction is the voltage U'_{L2-L3}, which in relation to the phase-to-phase input voltage at the second input transformer (1b) is rotated by the fixed angle of α by the phase shifter (10). Only the current signal I'_{L1} is applied to the comparator under normal load conditions, because the voltage signal U'_{L2-L3} is interrupted by the solid-state switch (4). The voltage is only switched through to the comparator when the current exceeds the pick-up setting of the overcurrent detector (3), i.e. after a fault has been detected. The direction of fault energy is therefore only evaluated in the phase or phases involved in a fault. The phase comparator operates when the phase-angle γ between the variables I'_{L1} and U'_{L2-L3} exceeds 90°. This is the situation shown in Fig. 6.9b for a three-phase fault in the direction of the line. The straight line FG is the limit line dividing the operating and restraint areas, i.e. the vector I'_{L1} is above the line FG for faults in the restraint direction and below it for faults in the operating direction. If the input current I_{L1} lies on the line M the angle γ would equal 180°, because for this case the vector I'_{L1} is on the line FG.

To enable the protection to detect solid three-phase faults close to the relay location although the system voltages may be almost zero and therefore directional measurement by the phase comparator impossible, the scheme also includes an undervoltage unit (5). The latter produces an output signal when the input voltage falls below 0.1 V and enables tripping after the 45 ms delay of the timer (9) without regard to the decision by the phase comparator (11). As an alternative to the indiscriminate tripping by the undervoltage unit, a "memory" feature - basically a resonant circuit for sustaining the voltage signal long enough for measurement - can be added to the voltage circuit of the directional relay.

6.2 Current differential protection

The general principle of current differential protection was described for the example of a two-winding power transformer in Section 3.2, where the difference between the currents entering and leaving a power transformer is monitored (see Fig. 3.1 and equation (3.2)). The difference between the currents entering and leaving a generator, an HV motor or any other item of power system plant can be compared in the same way. In all these cases, however, the basic required of differential protection must always be fulfilled, i.e. the protection must restrain for faults outside the zone of protection and must trip for faults inside the zone of protection. Devising a scheme which will always detect internal faults does not present any significant problems, but special precautions have to be taken to pre-

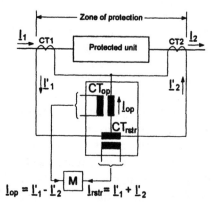

Fig. 6.10: Measuring principle of a stabilized differential protection (percent differential protection

$\quad$ CT$_{op}$ $\quad$ = interposing c.t. in the operating circuit

$\quad$ CT$_{rstr}$ $\quad$ = interposing c.t. in the restraint circuit

$\quad$ M $\qquad$ = amplitude comparator

vent the scheme shown in Fig. 3.1 from false tripping during 'through' faults. A method of restraining the protection is illustrated in Fig. 6.10. It can be seen that compared with the scheme without restraint in Fig. 3.1, a second variable is monitored which under normal load conditions and during external faults is the sum of the secondary currents I'_1 and I'_2, i.e.

$$I_{rstr} = I'_1 + I'_2 \tag{6.4}$$

The differential (operating) current under the same conditions is given by

$$I_{op} = I'_1 - I'_2 \tag{6.5}$$

The two variables I_{op} and I_{rstr} are compared in the amplitude comparator M and since in this case I_{rstr} is much greater than I_{op}, the protection does not trip.

For a fault in the protected zone which is fed from both sides, the above equations change to

$$I_{op} = I'_1 + I'_2$$
$$\tag{6.6}$$
$$I_{rstr} = I'_1 - I'_2$$

and since now I_{op} is much greater than I_{rstr}, the protection trips. If fault energy is only able to flow from one side, one of the secondary currents, e.g. I'_2, becomes zero. This would mean that $I_{op} = I_{rstr} = I'_1$ and tripping would not take place. For this reason, the restraint current is reduced by a factor of $k_{rstr} < 1$ so that $I_{op} > k_{rstr} \cdot I_{rstr}$ and the protection trips. The factor k_{rstr} is referred to as the pick-up ratio of the differential protection and typically has a setting range of 0.2 to 0.5. A setting is also provided for the basic setting I_g which determines the sensitivity of the protection at low currents.

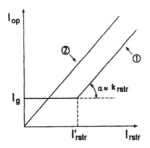

Fig. 6.11:   Operating characteristic of biased differential protection ① and the ratio of I_{op} to I_{rstr} for internal faults fed from one side ②

Figure 6.11 shows the operating characteristic of a solid-state biased differential protection. The operating region lies above the line ① and the restraint region below it. The line ② represents the ratio of I_{op} to I_{rstr} for internal faults fed from one side.

The principle of amplitude comparison employed in a solid-state differential relay can be seen from Fig. 6.12 [6.4]. The input variables I_{op} and I_{rstr} are supplied by the interposing c.t's in Fig. 6.10. The operating current I_{op} is applied to the bandpass filter F_i which is tuned to the rated frequency of the power system and attenuates all other frequencies. The current at the output of the filter is rectified by the bridge rectifier G_{op} and produces a proportional voltage drop U_{op} across the burden R_{op}. The restraint current I_{rstr} is similarly rectified by the bridge rectifier G_{rstr} and also produces a proportional voltage drop U_{rstr} across the burden R_{rstr}. The voltage U_{rstr} exerts a non-linear influence on the operation of the relay and causes the bend in the characteristic at a given restraint current I'_{rstr} (see Fig. 6.11) due the inclusion of the diode D1 and resistor R1. The basic pick-up current I_g is set with the aid of the resistor R2.

When a biased differential scheme is applied to protecting a power transformer, additional precautions are necessary to prevent false tripping when the power transformer is energised off-load. Depending on the instant at which the circuit-breaker is closed (worst case is the zero-crossing of the voltage), the initial (inrush) current can reach values several times the rated current of the transformer. Only after the transient current has decayed does the magnetization current reduce to its steady-state value of a few percent of the transformer rated current. Since the relatively high inrush current only flows in the winding on the supply side and the secondary current equals zero, the conditions seen by the differential relay are virtually identical to those of an internal fault supplied from one side and the protection would trip unless something is done to prevent it.

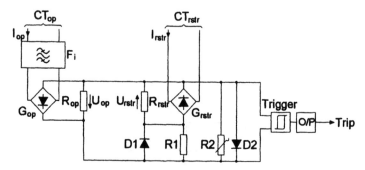

Fig. 6.12:   Amplitude comparison of the operating current I_{op} and restraint current I_{rstr} in biased differential protection [6.4]

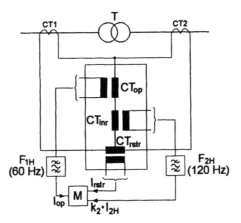

Fig. 6.13:   Differential protection scheme for a two-winding power transformer with second harmonic restraint for inrush currents

F_{1H}, F_{2H} = bandpass filters for first and second harmonics

Thus a criterion is necessary which is capable of distinguishing between an inrush of magnetizing current and an internal fault. Modern differential relays rely for this criterion on the fact that the second harmonic component in the inrush current amounts to 70 % of the fundamental, while it only reaches a maximum of 30 % in the fault current [6.5]. If therefore a significant second harmonic component is detected by a filter included for the purpose, the protection is restrained. The corresponding arrangement is shown in a simplified form in Fig. 6.13. Compared to Fig. 6.10, a second interposing c.t. is included in the spill current leg of the circulating current circuit which supplies the differential current to the bandpass filter F_{2H}. The resonant circuit of the filter amplifies any second harmonic and applies it to the restraint side of the amplitude comparator M.

Further details concerning the biased differential protection of power transformers and also the current differential protection of other plant (lines, busbars etc.) are discussed in Chapters 7 and 8.

6.3 Distance protection

6.3.1 General principle

Distance protection monitors the impedance of the protected unit as seen from the relay location (see Section 3.7). This impedance is high providing there is no fault on the protected unit and low when there is. Distance protection is primarily used to detect faults on transmission lines, but is also applied as back-up protection for large generators, power transformers and auto-transformers.

The most important advantage of distance protection in comparison with graded time-overcurrent protection as described above is its extremely fast operation for faults in the first zone of protection, which usually extends to between 85 and 90 % of section of line being protected (see Fig. 6.14). Thus faults close to the source can be tripped just as quickly as faults at the end of the line. From Fig. 6.14 it can also be seen that each of the distance relays RZ1 to RZ3 has several impedance zones with different operating times. The second and higher zones provide back-up protection for the other sections of line should either their protection or circuit-breaker fail to clear a fault. This means that assuming a fault on line L2 (fault location F2), distance relay RZ1 would trip circuit-breaker CB1 after the time $t_{2\mathrm{II}}$ should relay RZ2 fail to clear the fault in its first time step (operating time $t_{2\mathrm{I}}$). A fault at F3 (last 10 to 15 % of the line) on the other hand, would be detected by distance relay RZ2 in its second impedance zone and tripped after the time $t_{2\mathrm{II}}$. Should this not take place, RZ1 would not trip CB1 until the time $t_{3\mathrm{III}}$ had expired, because the fault lies in its third impedance zone.

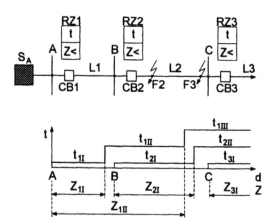

Fig. 6.14:   Grading of distance relays in a radial power system

It follows from the this explanation that discrimination is achieved in the case of distance protection by giving tripping priority to the relay which measures the lowest impedance, i.e. the shortest distance to the fault. It was also explained, however, that the first zone cannot be set to cover the full length of the line. In order to ensure proper discrimination under all conditions, it has to be set short and as a result faults in the last 10 to 15 % are detected in the second zone. This margin has to be observed, because if the first zone were to be set to 100 % of the line, faults at the beginning of the next section of line might be detected by mistake due to the inaccuracies of the distance measurement and line impedance data. This apparent disadvantage is overcome either by installing a communications channel between the relays at the two ends of the line or, if auto-reclosure is used, by an appropriate auto-reclosure logic. These possibilities will be discussed later.

A simplified illustration of the principle of distance measurement is given in Fig. 6.15. It assumes that only the positive-sequence component of the line impedance is measured and the zone of protection corresponds to the first zone of any of the relays, e.g. RZ1, shown in Fig. 6.14. This means that all faults occurring between station A and the zone limit G are in the operating area of the distance relay characteristic (Fig. 6.15a) and faults outside this area are not detected

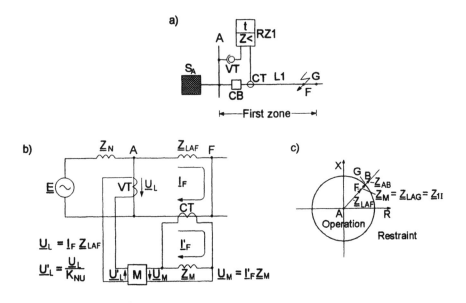

Fig. 6.15:   Fault detection by a distance relay on a radial line
a) protected unit   b) measuring principle   c) impedance characteristic

in the first zone. It is the relay's job to decide with the fault impedance lies within the characteristic or outside it. One possibility of doing this is to compare the absolute values of two variables, one of which is the voltage drop $\underline{U}_L$ across the impedance of the fault loop $\underline{Z}_{LAF}$ (Fig. 6.15b) and the other is the voltage drop $\underline{U}_M$ across what is referred to as the replica impedance $\underline{Z}_M$ connected to the secondary of the c.t. The reflected impedance of the replica on the primary side of the instrument transformers corresponds to the line impedance Z_{1I}, i.e. the impedance of the first zone. The two variables at the input of the amplitude comparator M are thus given by the relationships

$$\underline{U}'_L = \frac{\underline{U}_L}{K_{NU}} = \frac{\underline{I}_K\,\underline{Z}_{LAF}}{K_{NU}} \tag{6.7}$$

$$\underline{U}_M = \underline{I}'_K\,\underline{Z}_M = \frac{\underline{I}_K\,\underline{Z}_M}{K_{NI}} \tag{6.8}$$

where

$\underline{I}_K$ = primary fault current

$\underline{I}'_K$ = secondary value of the primary fault current $\underline{I}_K$, i.e. $\underline{I}'_K = \dfrac{\underline{I}_K}{K_{NI}}$

K_{NI} = nominal c.t. ratio
K_{NU} = nominal v.t. ratio
$\underline{Z}_{LAF}$ = impedance of the line loop between station A and the fault location F
$\underline{Z}_M$ = replica impedance, in this case $\underline{Z}_M = \underline{Z}_{1I}$

As mentioned above, the absolute values of the voltages $\underline{U}'_L$ and $\underline{U}_M$ are compared and therefore for a fault at the limit of the first zone (G in Fig. 6.15a)

$$|\underline{U}'_L| = |\underline{U}_M| \tag{6.9}$$

The means that the signals applied to the amplitude comparator M are balanced and the condition for tripping is not fulfilled. If both sides of equations (6.7) and (6.8) are divided by $\underline{I}_K$ and $\underline{I}'_K$, equation (6.9) becomes

$$|\underline{Z}'_L| = |\underline{Z}_M| = |\underline{Z}_{1I}| \tag{6.10}$$

The graphical representation of the characteristic in the R/X plane (see Section 3.7) is a circle with its center at the origin of the system of coordinates (Fig. 6.15c). This is also the location of the beginning of the protected line. Since the ratio between X and R of the line is known, the impedance of the line between stations A and B can be represented by the impedance vector $\underline{Z}_{AB}$ if a corresponding scalar value is assumed. The operating area of the protection lies inside the circular characteristic and the restraint area outside. Thus the tripping condition is expressed by the following inequality

$$|\underline{Z}_M| > |\underline{Z}'_L| \qquad (6.11)$$

which in terms of the voltages means $|\underline{U}'_L| < |\underline{U}_M|$ and is the case assumed in Fig. 6.15. The fault location F lies in the first zone and therefore $|\underline{Z}_{LAF}|$ is less than $Z_{LAG} = Z_M$. Since the same fault current flows through both impedances, the voltage $|\underline{U}'_L|$ is also less than $|\underline{U}_M|$ and the protection trips.

For faults beyond the limit of the first zone, $|\underline{Z}'_L| > |\underline{Z}_M|$ and therefore also $|\underline{U}'_L| > |\underline{U}_M|$ and the protection restrains.

The considerations up to the present were concerned with the application of distance protection to radial lines. If fault energy can be supplied to a line from both ends, distance relays have to be installed in both terminal stations as shown in Fig. 16.6a. Assuming that the relays should only trip for faults in the forwards direction, the two relays must have operating characteristics which are capable of making a clear directional decision. A protection with an impedance characteristic according to Fig. 6.15a cannot fulfil this requirement, because the impedance vectors seen by relays RZ1 and RZ2 for the fault locations F1 and F3 (Fig. 16.6a) are also in the operating area. For the respective relays, these faults are in the fourth quadrant of the R/X plane as shown for RZ1 in Fig. 6.16c. To prevent a distance relay from operating for faults in the reverse direction, its characteristic in the R/X plane must not encroach on the third quadrant. The desired mode

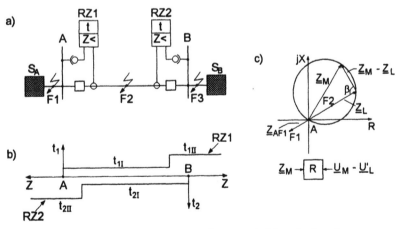

Fig. 6.16: Distance protection applied to a line supplied from both ends
a) protected line   b) impedance/time-step diagram   c) 'mho' characteristic for the relay RZ1

of operation is achieved by relays having a 'mho' characteristic (mho = inverse of ohm) as shown in Fig. 6.16c. A characteristic of this kind can be obtained using both amplitude and phase comparators. In the case of a phase comparator, the relevant input signals are

$$\underline{S}_1 = \underline{U}'_L \quad \text{and} \quad \underline{S}_2 = \underline{U}_M - \underline{U}'_L \tag{6.12}$$

from which can be derived

$$\underline{S}_1 = \underline{Z}'_L \quad \text{and} \quad \underline{S}_2 = \underline{Z}_M - \underline{Z}'_L \tag{6.13}$$

The parameter monitored by the phase comparator is the phase-angle β between the input signals (Fig. 6.16c). The limit value of $\beta = 90°$ for this angle describes the circular operating characteristic, i.e. $\beta < 90°$ for faults outside the first zone and $\beta > 90°$ for faults inside the first zone.

Apart from the circular mho characteristic shown above to illustrate the basic principle of distance protection, operating characteristics with shapes more ideally suited to the conditions in practice are achieved by solid-state distance relays. These and the measuring systems for obtaining them will be discussed later in this Section.

6.3.2   Types of distance relays

According to the number of measuring systems included, distance relays are divided into two main groups:
- distance relays with a single measuring system
- distance relays with several measuring systems

6.3.2.1 Distance relays with a single measuring system

Distance relays of this kind are used primarily where the permissible time for detecting faults in the first zone is not extremely short which generally applies in MV and HV power systems. The relatively slow operating time for the first zone is in the region of tens of milliseconds. The relays take so long, because the voltages and currents needed to determine fault distance and direction have to be selected and applied to the measuring system. This function is performed for the different kinds of faults by fault detectors or starting units.

The basic structure of a distance relay with a single (switched) measuring system can be seen from Fig. 6.17 [6.7]...[6.12]. The signals corresponding to the c.t. secondary currents are pre-processed in the signal conditioning unit 1a, respectively the signals derived from the v.t. secondaries in the signal conditioning unit 1b. The signal conditioning units adjust the signal levels for processing by the solid-state circuits and screen or filter out any interference. The signals then go to the fault detectors 2 and selection system 3. There are several

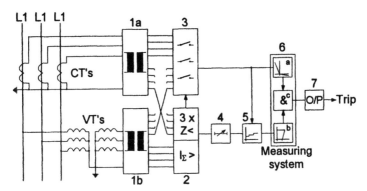

Fig. 6.17:   Basic configuration of a distance relays with a single measuring system [6.10]

fault detectors - in this example underimpedance units - which pick up according to the kind of fault on the power system, i.e. the phases involved in the fault. When one or several fault detectors pick up, the timer 4 is started and control signals are passed to the selection logic 3, which applies the corresponding current and voltages for determining fault direction to the measuring unit 6. In order to determine fault distance, the same signals reach the measuring unit via the range switching unit 5 which controls the impedance reach of the relay in relation to the time-step characteristic of the timer 4. The operating characteristic of the relay for the various impedance zones is obtained by combining the output signals of the directional unit 6a and the distance unit 6b in the AND gate 6c. A fault lying inside one of these characteristics causes a signal to be applied to the output unit 7 which then trips the protected line.

In the example given above, underimpedance units are used for the fault detectors or starters and the measuring unit has a so-called polygon characteristic. The are, however, many alternative shapes of characteristics, each of which offers advantages depending on the fault conditions in the system, type of system grounding, length and load level of the line to be protected etc. These alternatives will be described in detail below.

6.3.2.2 Distance relays with several measuring systems

Distance relays with several measuring systems are used mainly in EHV systems, but are also to be found in many HV systems. Their most important advantages [6.13] are
- very short operating times for first zone faults, because the correct currents and voltages are permanently connected to the relevant measuring system and do not have to be selected after the fault occurs.
- specific measuring systems, thus permitting characteristics to be optimized for phase and ground faults
- high reliability of measurement during evolving faults, i.e. when the type of fault changes progressively

The multi-system distance relays most frequently encountered have three, four or six measuring systems [6.13...6.17]. Figure 6.18 shows the block diagram of a distance relay with six measuring systems per time step. Three measuring systems per time step are for detecting ground faults and three for phase-to-phase faults. A relay of this kind will have two or three of these groups depending on whether it has two or three time steps. The measuring systems for the maximum reach can also serve as starting units for the timers etc. Starters of

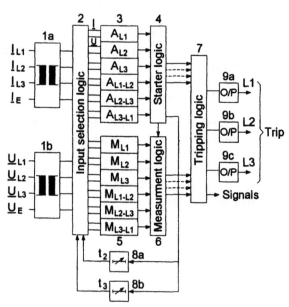

Fig. 6.18   Block diagram of a distance relay with six measuring systems

the type used in a relay with a single measuring system are not strictly necessary, but are usually still included to perform other functions (Section 6.3.3).

The signals coming from the c.t's and v.t's are pre-processed as before in the signal conditioning units 1a and 1b, which adjust the signal levels for processing by the solid-state circuits. The outputs of the input variable selection logic 2 which also includes the replica impedance voltages, apply the correct signals for measurement to the fault detectors 3 and the measuring systems 5. The starting logic 4 derives the various detection criteria for starting any timers 8a and 8b which may be fitted and enabling the tripping logic 7 from the input signals. The measuring logic 6 performs the task of monitoring the impulses coming from the measuring systems 5 and deriving the appropriate continuous signals required by the tripping logic. The latter controls the tripping units 9a, 9b and 9c either independently for single-phase tripping or together for three-phase tripping of the circuit-breaker poles in the phases L1, L2 and L3.

6.3.3   Starting criteria and starting systems

As stated above, fault detectors or starting systems are used in distance relays with a single measuring system as well as in relays with several measuring systems. The main purpose of the starters in distance relays with a single or with just three measuring systems is to apply the correct currents and voltages according to the kind of fault to the measuring system or systems. This is no longer necessary in distance relays with four or six measuring systems. Nevertheless they are still included, amongst other things to prevent false tripping during balancing phenomena (e.g. following switching operations) which take place without there being a fault on the system [6.18]. Other tasks performed by the starters are:
- tripping signal selection in single-phase auto-reclosure schemes
- reversal of measuring direction at a definable time after fault incidence
- enabling tripping by other functional units of the distance relay
- redundant back-up protection for the measuring systems [6.6]

Starting units can be of the following types:
- overcurrent (only in MV power systems)
- angle-insensitive underimpedance
- angle-sensitive underimpedance

Overcurrent starters are used primarily in MV power systems, because the minimum fault current for a fault at the maximum reach of the protection is sufficiently higher than the maximum load current. The number of starters installed in a relay depends on the method of system grounding. Starters monitor the currents of all three phases in solidly grounded systems and the currents of phases L1 and L2 and the summation current (neutral current) for the detection of cross-country faults in systems with Petersen coils. The block diagram of a dis-

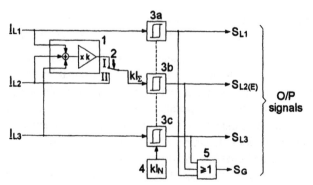

Fig. 6.19:   Block diagram of the overcurrent starters in a distance relay [6.9]

tance relay with overcurrent starters is shown in Fig. 6.19 [6.9]. In this particular version, overcurrent starting unit 3b in the center can be switched to measure either the current of phase L2 or the summation current I_Σ, which is derived in 1 and amplified by the factor k (usually k = 2). The selector switch 2 determines which system variable is monitored by the overcurrent unit 3b. The switch is in position I for ungrounded systems and systems equipped with Petersen coils and position II for solidly grounded systems. The three overcurrent starters 3a to 3c are set by the ganged decade switch 4. The input selection logic is controlled by the output signals S_{L1}, $S_{L2(E)}$ and S_{L3}. A common starting signal S_G is obtained via the OR gate 5.

Angle-insensitive underimpedance starters are used wherever the minimum fault current is less than the maximum load current and therefore the application of overcurrent starters is impossible. Underimpedance measurement implies that the fault voltage as well as the current have to be evaluated. The voltage can either be included directly to perform a "genuine" underimpedance measurement or it can be used indirectly.

Genuine underimpedance starters can be recognised from the fact that their reach is set in Ω/phase and is thus clearly defined. This is achieved by appropriate selection of the replica reactance X_M in 1 of Fig. 6.20. The difference voltage $\underline{U}_1$ and the summation voltage $\underline{U}_2$ are derived by 2a and 2b from the input variables $\underline{U}_F$ and $\underline{I}_F$ measured for the phase with the fault, $\underline{I}_F$ being first converted in 1 to the voltage drop $\underline{I}_F \cdot X_M$. $\underline{U}_1$ and $\underline{U}_2$ are then applied to the phase comparator 3 which produces a signal at its output when the angle α between the voltages exceeds 90°. As an additional safety measure to prevent transients from causing incorrect operation of the starters, the output is enabled by an overcurrent interlock 4 via the AND gate 5. The overcurrent interlock is normally set to pick up at $0.2\ I_N$ where I_N is the rated current of the distance relay.

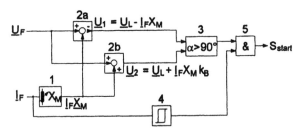

Fig. 6.20:   Block diagram of an underimpedance starting unit [6.7]

The corresponding operating characteristic in the complex and R/X planes is shown in Fig. 6.21. The impedance vectors result when the voltage vectors are divided by $\underline{I}_F$.

Figure 6.22 shows a three-phase starting system with three underimpedance units 2 and an overcurrent unit 4 for detecting a neutral current. The voltage inputs of the underimpedance starters are normally connected to a phase-to-phase potential and are switched over to a phase-to-neutral potential if the overcurrent unit picks up. This system is used in distance relays with a single measuring unit. The reach of the underimpedance units is set on the decade switch 3 and the pick-up current of the overcurrent unit on the decade switch 5. The output signals S_{L1}, S_{L2}, S_{L3}, S_E and S_G control the selection of the current and voltages used for measurement and start the timing unit.

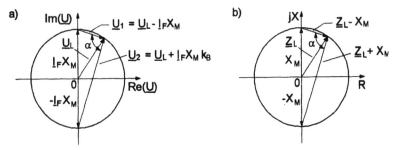

Fig. 6.21:   Operating characteristic of an angle-insensitive underimpedance starting unit in the complex plane a) and R/X plane b)

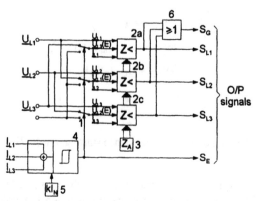

Fig. 6.22:   Underimpedance starting system for a distance relay with a single measuring unit [6.7]
1 = solid-state switch

An alternative type of angle-insensitive underimpedance starter is basically a voltage controlled overcurrent unit comprising an undervoltage relay (U<) and an overcurrent relay (I>). As can be seen from characteristic shown in Fig. 6.23a, starting can only take place when the voltage falls below the setting U_{set} at the same time as the current exceeds the setting I_{set}. Typical setting ranges are 0.5 ... 0.95 U_N and 0.25 ... 1 I_N [6.10, 6.18]. The underimpedance characteristic in this case is cut off by a second, high-set overcurrent unit (I>>). This indirect principle of monitoring impedance results in an Z/I characteristic which can be ideally adapted to power system requirements at moderate fault levels (Fig.

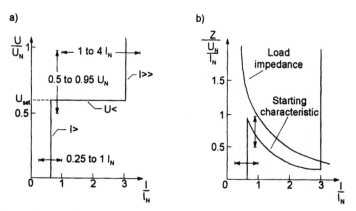

Fig. 23:   Characteristic of a voltage controlled overcurrent starter [6.18]
a) U/I characteristic   b) Z/I characteristic

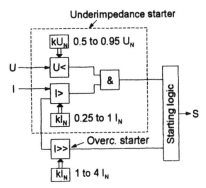

Fig. 6.24:   Combined operation of an undervoltage and two overcurrent units to produced the characteristics of Fig. 23 [6.10]

6.23b). A block diagram illustrating the combined operation of the three starting
elements is given in Fig. 6.24.

Angle-sensitive underimpedance starters were especially developed for
situations where fault currents are low and the voltage drop produced by a fault
at the relay location is relatively small. Such conditions prevail on long heavily
loaded lines supplied by stations with a high short-circuit power and also railway
system feeders. Faults are detected by monitoring the phase-angle between the
current and voltage of a conductor, which is larger during a fault than under
normal load conditions. Either direct or indirect impedance measurement can be
employed. One possible solution for the direct type of measurement is the same
as that shown in Fig. 6.20 with the exception that the operating angle α of the
phase comparator 3 is, for example, 130° instead of 90° [6.15], which produces
the lenticular characteristic similar to Fig. 6.25. The reach in the forwards direction Z_A is set by varying the amplification of the fault voltage component $\underline{U}_F$ in
the difference voltage $\underline{U}_1$. The reach in the reverse direction is a function of the
factor k_B by which the replica voltage $I_F Z_M$ is reduced when deriving the sum-

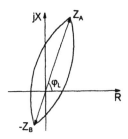

Fig. 6.25:   Lenticular starting characteristic

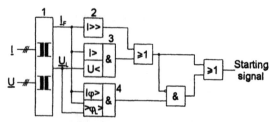

Fig. 6.26:   Combined operation of the various units of an indirect angle-sensitive underimpedance starter [6.19]

mation voltage $\underline{U}_2$. The two voltages applied to the phase comparator are thus

Difference voltage:

$$\underline{U}_1 = \underline{U}_F - I_F Z_M \tag{6.14}$$

Summation voltage:

$$\underline{U}_2 = \underline{U}_F + k_B I_F Z_M \tag{6.15}$$

Indirect angle-sensitive underimpedance starting is a combination of phase comparators and overcurrent and undervoltage units as shown in Fig. 6.26 [6.19]. The high-set overcurrent unit 2 and the voltage controlled overcurrent starter 3 are identical to those described above. The phase-angle control unit 4 comprises an overcurrent unit $I_\varphi>$ with a setting range of 0.25 ... 1 I_N and a phase comparator with a fixed setting between 50° and 100°. This range covers the possible phase-angles φ_F between the fault current I_F and the fault voltage $\underline{U}_F$. It can be seen from Fig. 6.26 that the phase-angle control becomes ineffective if either the high-set overcurrent or the voltage controlled overcurrent unit picks up. The operating area of the phase-angle controlled starter according to

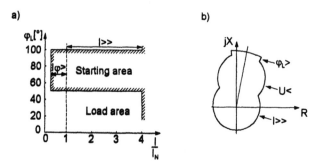

Fig. 6.27:   Operating area of a phase-angle controlled starter a) and the corresponding characteristic in the R/X plane b) [6.14]

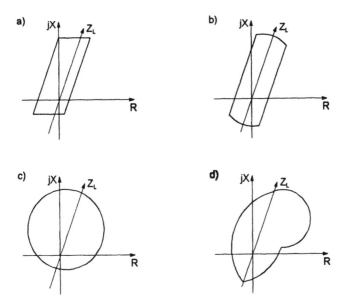

Fig. 6.28:   Examples of angle-sensitive starting characteristics for distance relays
a) off-set polygon   b) oval   c) off-set mho   d) lens/circle hybrid

Fig. 6.26 in relation to the relative fault current is given in Fig. 27a and the corresponding characteristic in the R/X plane in Fig. 27b [6.14].

Apart from those described above, other angle-sensitive starters with a variety of characteristics are also in use. The R/X diagrams of some of these are given in Fig. 6.28.

6.3.4   Measuring systems for determining fault distance and direction

The majority of modern solid-state distance relays employ phase comparators for determining the distance and direction of a fault. According to the type of characteristic, either a single unit performs both functions or distance and direction are measured by separate units. A typical example for the characteristic and arrangement of a directional distance measuring system is shown in Fig. 6.29a and b. The phase comparator PC monitors the angle α between the voltages $\underline{U}_1$ and $\underline{U}_2$ and operates when it exceeds $90°$. The difference voltage $\underline{U}_1$ must fulfil the following conditions to enable both three-phase faults and ground faults to be detected

$$\underline{U}_1 = \underline{U}_F - \underline{U}_M \qquad\qquad (6.16)$$

where

$$\underline{U}_M = (\underline{I}_F + \underline{k}_0\underline{I}_0)\underline{Z}_M \qquad (6.17)$$

In the above equations, $\underline{I}_F$ and $\underline{U}_F$ are the current and voltage of the fault loop, $\underline{Z}_M$ the replica impedance, $\underline{I}_0$ the neutral current and $\underline{k}_0$ the zero-sequence current factor set on the distance relay and given by the ratio $\underline{k}_0 = (\underline{Z}_0 - \underline{Z}_1)/3\underline{Z}_1$, where $\underline{Z}_1$ and $\underline{Z}_0$ are the positive and zero-sequence impedances of the protected line.

In the case of the off-set mho characteristic shown in Fig. 6.29a, the second voltage used for phase comparison $\underline{U}_2$ (also referred to as the reference voltage $\underline{U}_{ref}$) is

$$\underline{U}_2 = \underline{U}_{ref} = a\underline{U}_F + b\underline{U}_H \qquad (6.18)$$

where

$\underline{U}_H$ = healthy phase-to-phase voltage

a, b = factors or proportionality between 0 and 1

For a = 1 and b = 0, $\underline{U}_2$ equals the fault voltage $\underline{U}_F$ and the characteristic becomes a mho circle in the R/X plane (see Fig. 6.16c) which is independent of the source impedance $\underline{Z}_S$. The particular disadvantages a pure mho characteristic [6.20] are
- limitations on detecting faults which include arc resistance especially for close faults and faults on short lines
- inadequate directional sensitivity for very low fault voltages

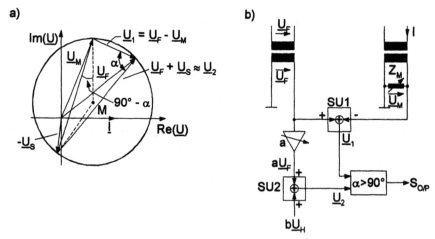

Fig. 6.29: Off-set mho characteristic a) and corresponding measuring system b) [6.20]

It is for this reason that the voltage $\underline{U}_2$ includes a portion of a healthy system voltage $\underline{U}_H$ which transforms the operating characteristic for faults in the forwards direction to the off-set mho shown in Fig. 6.29a. Depending on the value of the source impedance, the introduction of a healthy system voltage shifts the mho circle to a greater or lesser extent in the direction of the real axis and thus achieves a much improved accommodation of high-resistance faults.

In the case of close, solid, three-phase faults, there is neither a healthy system voltage nor sufficient fault voltage to ensure a correct directional decision and for this reason a so-called memory feature is included, which is basically a resonant circuit for sustaining the reference voltage for a few periods after the occurrence of a three-phase fault [6.21].

Apart from the combined directional distance measuring system using a single phase comparator described above, the tasks of determining distance and direction can also be performed separately. While such measuring systems require several phase comparators, they do have the advantages of being able to define the operating area in the R/X plane independently of the source impedance $\underline{Z}_s$ and of adequately accommodating arc resistance even on short lines. Typical of such systems is the polygon operating characteristic of Fig. 6.30 as constructed by the measuring system of Fig. 6.31 which superimposes the distance characteristic D1/D2 on the directional characteristic R1/R2 [6.19, 6.22].

Two voltages are compared to determine fault distance (Fig. 6.31)

$$\underline{U}_1 = \underline{U}_F - \underline{U}_i$$
$$\underline{U}_2 = \underline{U}_i \tag{6.19}$$

where

$$\underline{U}_i = I_F \cdot R_{sh} \cdot e^{j\vartheta} \tag{6.20}$$

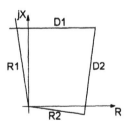

Fig. 6.30: Polygon characteristic of measuring systems in modern distance relays
D1, D2 = limit lines of the distance unit
R1, R2 = limit lines of the directional unit

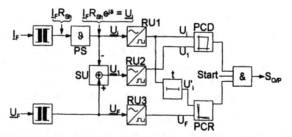

Fig. 6.31:   Block diagram of a measuring system with a polygon characteristic [6.19]
             PCD = phase comparator for determining distance
             PCR = phase comparator for determining direction
             PS = phase-shifter

U_i is thus proportional to the voltage drop produced by the fault current I_F across a shunt R_{sh} advanced by a phase-angle ϑ. Figure 6.32 illustrates the construction of the distance characteristic and the phase-angles of the corresponding variables for internal (Fig. 6.32a) and external (Fig. 6.32b) faults. The two limit lines D1 and D2 are shifted by an angle ϑ in relation to U_i. The comparator monitors the angle α between the variables U_1 and U_2 as defined by equation (6.19). The operating point for the distance measurement is given by the limit angle $\alpha_1 = \pi - \alpha$. The angle α is always greater than α_1 for faults within the area bounded by the line D1 and D2 and less than α_1 for faults outside these limits. The limit lines D1 and D2 thus determine the reach of the measurement in the forwards direction.

The directional characteristic is similarly derived by phase-comparator PCR (Fig. 6.31) from the voltages U_F and U'_i. U_F in this case represents the fault

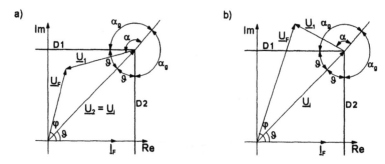

Fig. 6.32:   Construction of the distance characteristic by the distance measurement
             arrangement according to Fig. 6.31 [6.19]
             a) internal faults, i.e. $\alpha > \alpha_g$ b) external faults, i.e. $\alpha < \alpha_g$

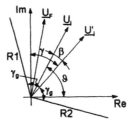

Fig. 6.33:   Construction of the directional characteristic in a distance relay according
to Fig. 6.31 [6.19]

voltage after it has been shaped into a squarewave and $\underline{U}'_i$ is a squarewave voltage obtained by shaping and phase-shifting $\underline{U}_i$ in the timer t, the voltage $\underline{U}_i$ being
proportional to the current. The relationships between the variables and the deri-
vation of the limit lines R1 and R2 for the determination of fault direction can be
seen from Fig. 6.33. The positions of R1 and R2 are determined by the limit an-
gle γ_l which is defined in relation to the voltage vector $\underline{U}'_i = \underline{U}_i \cdot e^{j\vartheta}$. For faults in
the forwards direction, the monitored angle γ between the voltages $\underline{U}_F$ and $\underline{U}'_i$ is
less than γ_l and greater for faults in the reverse direction. To prevent the directional decision from being influenced by the fault voltage $\underline{U}_F$, another voltage is
used instead in many instances.

An alternative method of arriving at a polygon operating characteristic uses
three independent phase comparators [6.23], one each for limiting the reach in
the reactive and resistive directions (distance measurement) and one for
determining fault direction. The principle is show in Fig. 6.34.

The reactive limit line for directional measurement is generated by the phase
comparator PCD which monitors the angle γ_1 between the variables $\underline{U}'_M$ and
$\underline{U}_{1D}$, where

$$\underline{U}'_M = \underline{I}_F \cdot \underline{Z}_M \cdot e^{j\delta}$$

$$\underline{U}_{1D} = \underline{U}_F - \underline{I}_F \cdot \underline{Z}_M$$

(6.21)

where $\underline{Z}_M$ is the replica impedance. The comparator operates when the angle γ_1
exceeds 90°. This determines the position of the reactance line D which limits the
reach in the forwards direction.

The operating area in the direction of the real axis is bounded by the resistive
limit line W which is determined by the limit set for the phase-angle γ_2 between

the input variables $A\underline{U}_M$ and $\underline{U}_{1W}$ of the phase comparator PCW, where

$$A\underline{U}_M = A\underline{I}_F \cdot \underline{Z}_M$$

$$\underline{U}_{1W} = \underline{U}_F - A\underline{U}_M \qquad (6.22)$$

and A is the amplification factor. The limit for γ_2 is either 28° or 152° according to whether $\underline{U}_{1W}$ is leading or lagging in relation to $\underline{U}_M$.

a)   b)

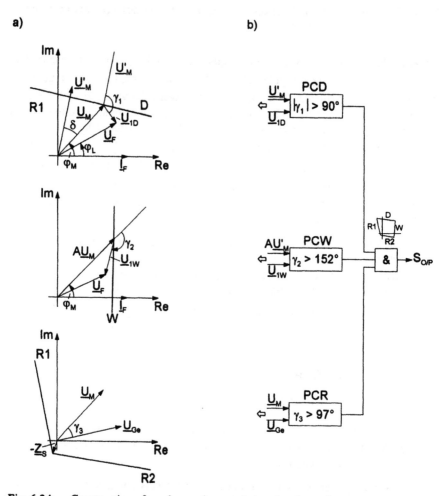

Fig. 6.34: Construction of a polygon characteristic using three phase comparators [6.23]
a) individual characteristics   b) signal comparison and logic

Phase comparator PCR monitors the angle γ_3 between the replica voltage $\underline{U}_M$ and a phase-shifted healthy voltage $\underline{U}_H$. A combination of the fault voltage $\underline{U}_F$ and a voltage supplied by a memory feature is used in the case of a three-phase fault. The phase comparator operates when the angle γ_3 exceeds $97°$.

6.3.5   Resulting distance relay operating characteristics

By superimposing the starting characteristics described in Section 6.3.3 and the measuring system characteristics described in Section 6.3.4 for the various zones of protection (Fig. 6.14), a composite operating characteristic for a given distance relay can be represented in the R/X plane. Figure 6.35 shows two examples for the same line. The diagrams assume that only three stages having reaches determined by the impedances Z_I, Z_{II} and Z_{III} are used in the forwards direction. The overall reach of the protection is defined by the setting Z_{start} of the starters. Finally, it must be pointed out that only the basic types of distance protection scheme with a "normal" first zone setting of 85 % to 90 % of the line impedance

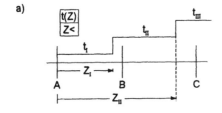

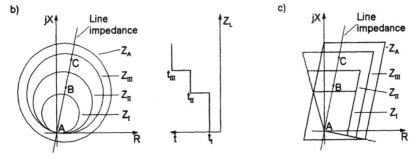

Fig. 6.35:   Composite distance relay characteristics
a) protected sections of line   b) circular characteristics   c) polygon characteristics

Z_{AB} have been dealt with up to the present. There are, however, other possibilities such as so-called acceleration schemes with a first zone extended to between 115 % and 130 % of Z_{AB} and also schemes with zones which measure in the reverse direction. These alternatives will be discussed in Chapter 7.

7

Analogue Protection for Power Systems

7.1 Line protection

7.1.1 Methods of protection

The kind of protection employed and the amount of protection installed for protecting cables and overhead lines depend mainly on the importance of the particular feeder and the type of power system grounding. The faults which can occur in practice thus vary, but generally fall into the categories phase faults, ground faults and overloads. Table 7.1 gives an overview of the protection techniques used to detect phase and ground faults in general and in Section 7.1.4.1 those for detecting ground faults in ungrounded systems and systems with Petersen coils. Only short heavily loaded lines are normally equipped with protection against overloads. Since, however, the principle of overload protection employed does not differ from one item of plant to another, it is not gone into further in this Section.

Table 7.1   Phase and ground fault protection in MV, HV and EHV power systems

Kind of fault	Protection device	
	MV lines	HV/EHV lines
Phase and ground faults	- overcurrent (delayed and instantaneous) - directional overcurrent - differential (pilot wire) - distance	- differential (pilot wire) - distance - signal comparison - phase comparison - directional comparison

7.1.2   Protection devices for detecting phase faults

7.1.2.1 Non-directional and directional overcurrent protection

The application of this kind of protection is restricted to MV power systems. Time-overcurrent protection is the main protection used on radial lines (Section 6.1.2) and is supplemented to maintain discrimination on double-circuit lines and ring lines by directional units (Section 6.1.4). Instantaneous overcurrent units may also be added, providing a line is long enough for them to detect phase-to-phase faults along at least 20 % of its length.

7.1.2.2 Differential (pilot wire) protection

The principle of a differential scheme involves the comparison of measured variables, i.e. the values of the currents - respectively voltages proportional to them - at the two ends of a transmission line are measured and compared. In contrast to the differential scheme for a protected unit of limited physical size (generator, power transformer etc.) described in Section 6.2, a differential scheme for a transmission line comprises two sets of protection equipment for a two-ended feeder and three for a teed feeder, one for each line terminal. A communication channel - generally pilot wires (hence the alternative name for this kind of protection) but also power line carrier and more recently optical fibers - is thus needed to transmit the signals from one end of the line to the other [7.1, 7.2].

Line differential protection using pilot wires can be of two types, a circulating current scheme or a balanced voltage scheme. Simplified illustrations of both arrangements can be seen in Fig. 7.1. An important difference between the two is that the circulating current scheme needs three pilot wire cores (Fig. 7.1a) and

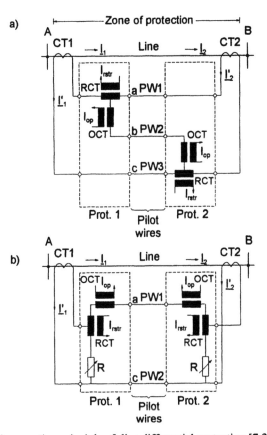

Fig. 7.1:   Basic operating principle of line differential protection [7.2]
             a) circulating current   b) balanced voltage

the balanced voltage scheme (Fig. 7.1b) only two. Both include through-fault stabilization based on the derivation of the restraint current I_{rstr}. During normal operation and through-faults, this current and therefore its proportional voltage are higher than the pick-up current I_{op}, while during an internal fault the relationship is the other way round, i.e. $I_{op} > I_{rstr}$, and the respective set of terminal equipment trips its line breaker.

Practical pilot wire schemes reduce the three phase currents at the line terminals by summation to a single-phase quantity in order to limit the number of pilot wire cores to two, respectively three (see Section 5.3.2.1). Different summation c.t. ratios are chosen for the three phases to prevent the fault current from canceling for particular faults. Such an arrangement is shown in Fig. 7.2 for a scheme

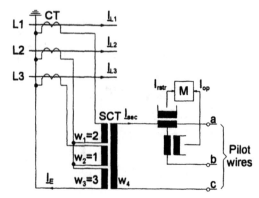

Fig. 7.2:   Summation c.t. for obtaining a single-phase quantity in a circulating current
             pilot wire scheme

according to Fig. 7.1a. The ratios in this example are w1 : w2 : w3 = 2 : 1 : 3,
so that the secondary current I_{sec} is given by

$$I_{sec} = 2I'_{L1} + I'_{L3} + 3I_E$$

where I_E is the residual current flowing in the star-point of the main c.t's which is
zero under normal load conditions and for faults not involving ground. As can be
seen from Fig. 7.1a, the pilot wires PW1 continuously conduct the currents I'_1
and I'_2 under normal load conditions and during through faults. This is not the
case for the second circuit of Fig. 7.1b where they only flow through the resistor
R and a part of the interposing c.t. ICT. The resistor R is adjusted to compensate
the loop resistance of the pilot wires PW1 and PW2 [7.5] which therefore link
points of equal potential under normal load conditions and no current flows.

The most important hazards which can adversely influence the operation of a
pilot wire protection scheme are

a) longitudinal voltages induced in the pilot wires during a ground fault on the
   protected line
b) charging currents of the pilot wire capacitances
c) pilot wire defects (short and open-circuit cores)
d) severe c.t. saturation due to the DC component in a fault current

High induced voltages can damage both the pilot wires themselves as well as
the protection equipment in the terminal stations. To overcome this disadvantage,
pilot wires with a high insulation level and isolating transformers between the pilot wires and the protection equipment are used.

Depending on the length of the pilot wires and the ground fault level, the pilot
wire charging currents during a through-fault can adversely affect protection

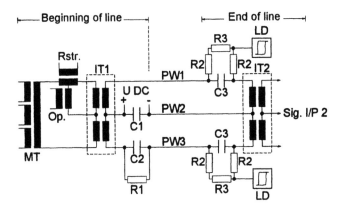

Fig. 7.3: Supervision of the three pilot wires of a pilot wire protection scheme

IT1, IT2 = isolating transformers

stability, i.e. the protection can mal-operate. A practical scheme must therefore include measures to compensate the capacitance of the pilot wires.

Disadvantages a) and b) above can be avoided by using an optical fibre link instead of conventional pilot wires. The data is normally exchanged serially in this case which necessitates the use of multiplexers in the terminal stations. On the other hand, the link can also be used to transmit data other than just the pilot wire information.

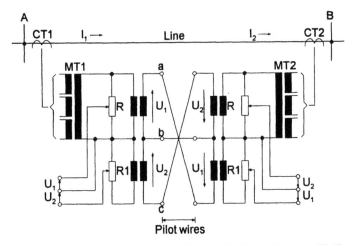

Fig. 7.4: Voltage comparison pilot wire scheme with three pilot wires [7.6]

The method of supervising the integrity of the pilot wires depends on the type of differential protection scheme. Figure 7.3 shows a supervision arrangement for a protection scheme using three pilot wires [7.7]. The DC injected at the beginning of the center core PW2 returns via cores PW1 and PW3. The resistors R1, R2 and R3, which are short-circuited for the AC signal transferred between the terminal stations by the condensers C2 and C3, bring about a potential difference along the pilot wires. This potential difference is detected in the event of an open or short-circuit by level detectors LD which then block the operation of the protection devices in the terminal stations.

What has come to be called the echo principle is the system generally used for supervising just two pilot wires. In this case, an AC at a frequency of about 1 kHz is injected instead of the DC [7.3, 7.5].

Where transient saturation of the c.t's occurs, the resulting severe distortion of the measured variables which takes place would cause false tripping of the protection during through-faults. To combat this situation, either two measurements in sequence are carried out [7.7] or the difference in time between the incidence of restraint and operating signals for an internal and an external fault is evaluated [7.6]. This principle is described in detail in connection with the differential protection of power transformers.

A different kind of line differential protection referred to as an opposed voltage scheme is shown in Fig. 7.4. Once again three pilot wires are used, but this time voltages U_1 and U_2 proportional to the currents in the terminal stations ($U_1 \cong I_1$ and $U_2 \cong I_2$) are compared [7.6]. The advantage of this scheme is that the pilot wires can be supervised without injecting a separate AC or DC. The voltage-drop and phase-shift caused by the pilot wire core resistance R and the capacitance C between the pilot wires are compensated in the relay input circuit

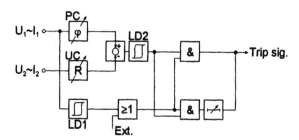

Fig. 7.5:   Relay measuring system used in the line differential protection of Fig. 7.4
             PC = phase compensation, UC = amplitude compensation,
             Ext. = external signal

as shown in Fig. 7.5. The level detector LD1 can be used either as an overcurrent enabling function or as a back-up time-overcurrent unit.

Because of the constraints of pilot wire resistance and capacitance, line differential protection is only suitable for relatively short lines (up to about 25 km). Protection schemes for longer lines employ other means of communication between the terminal stations and are usually of the phase comparison type.

7.1.2.3 Phase comparison protection

The operating principle of a phase comparison protection (PCP) scheme is based on monitoring the phase-angle between the currents measured at the two ends of a line (see Section 3.3 and Fig. 3.2). Either the phase currents are used for measurement or a composite current obtained from a mixing c.t. Phase comparison schemes thus fall into two categories, those which perform a three-phase comparison and those which perform a single-phase comparison. Both of these can then be divided into schemes which process both half-cycles, i.e. positive and negative, and those which process only one half-cycle [7.1].

Figure 7.6 shows the block diagram of a PCP scheme which measures the three phases individually [7.9], although only one phase is shown for reasons of simplification. The secondary current I_A of CT3 is applied to the input transformer IPT which provides electrical insulation between the secondary circuit of the main c.t. and electronic circuits of the protection and also adjusts the signal level for electronic measurement. Any interference and noise is eliminated by the bandpass filter BP1 such that the signal S_1 at the input of the shaper SH which is proportional to the phase current I_A is purely sinusoidal. The squarewave signal

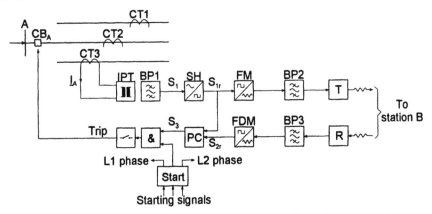

Fig. 7.6:   Block diagram for one line terminal of a three-phase phase comparison scheme

S_{1r} at the output of the shaper is then applied to the phase comparator PC and also to the frequency modulator FM which generates a corresponding AF signal. This signal passes via the bandpass filter BP2 to the AF transmitter T and thence to the communications channel equipment and an identical arrangement at the other end of the line. The same thing happens in the reverse direction with the audio frequency signal which is proportional to the phase current I_B from station B to the receiver R. This signal goes via the bandpass filter BP3 to the demodulator FDM which produces a squarewave signal S_{2r} at its output. The phase comparator PC compares the two squarewaves S_{1r} and S_{2r} and generates a signal S_3 when the phase-angle between the currents I_A and I_B is less than 180° - Θ, where Θ is the range of the blocking angle. However, the starting element Start must also pick up before a tripping signal appears at the output of the AND gate.

Figure 7.7 shows in a simplified form the principle of the phase comparison measurement performed by the protection for a through-fault a) and for a fault on the protected line b). The AF modulation is not shown and it is also assumed that only one half-cycle of the primary system current is monitored. It can be seen that for an idealized external fault, i.e. neglecting line charging and load cur-

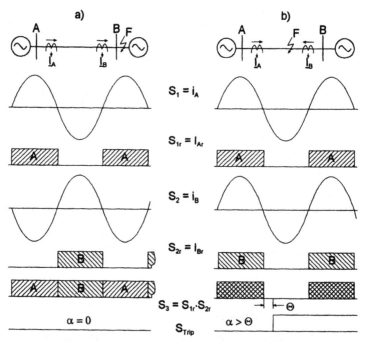

Fig. 7.7:   Phase comparison signals for an external fault a) and an internal fault b)

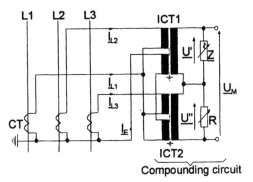

Fig. 7.8:   Compounding circuit for obtaining a single-phase voltage signal from three phase currents

rents, the comparator PC produces a steady-state signal corresponding to the product $S_{1r} \cdot S_{2r} = I_{Ar} \cdot I_{Br}$. There is no gap between the squarewaves derived from the positive half-cycles of the currents I_{Ar} and I_{Br} at the two ends of the line, i.e. $a = 0$. For a fault on the protected line under the same idealized conditions (Fig. 7.7b), there is a gap between the squarewaves S_{1r} and S_{2r} of 180° or 8.3 ms at $f_N = 60$ Hz and tripping takes place.

Where to economize on communication channels a three-phase measurement is reduced to a single-phase signal, the "mixed" current variable is a linear combination of the symmetrical components of the three phase currents. This can either be a combination of all the symmetrical components, i.e. $\underline{I}_1$, $\underline{I}_2$ and $\underline{I}_0$ or of only the positive and negative-sequence components. In the first case, the factors are chosen such that the mixed current becomes [7.13]

$$\underline{I}_M = \underline{I}_1 + 3\underline{I}_2 + 5\underline{I}_0 \tag{7.2}$$

and in the second case

$$\underline{I}_M = k_1\underline{I}_1 + k_2\underline{I}_2 \tag{7.3}$$

where typical values are $k_1 = \pm 1$ and $k_2 = 5...7$ [7.10]. Figure 7.8 shows a circuit for obtaining a single-phase signal voltage from the three phase currents, the output voltage U_M of which corresponds to the mixed current of equation (7.3) since

$$\underline{U}_M = K_D(\underline{U}' + \underline{U}'') = K_D\left[(\underline{I}_{L2} + \underline{I}_E)\underline{Z} + (\underline{I}_{L3} + \underline{I}_E)R\right] \tag{7.4}$$

$I_E = 0$ for a symmetrical three-phase load current and equation (7.4) becomes

$$\underline{U}_M = K_D(\underline{I}_{L2}\underline{Z} + \underline{I}_{L3}R) \tag{7.5}$$

where

K_D = ratio of the interposing c.t. ICT
$\underline{Z}$ = adjustable impedance burden of the compounding circuit
R = adjustable resistive burden of the compounding circuit

Taking the relationships between the phase currents and the symmetrical components into account

$$\underline{U}_M = K_D e^{j120°}\left(R + \underline{Z}e^{j120°}\right)\left(\underline{I}_1 + \underline{k}_2\underline{I}_2\right) \tag{7.6}$$

where

$$\underline{k}_2 = \left(\underline{Z} + Re^{j120°}\right)\big/\left(\underline{Z}e^{j120°} + R\right) \tag{7.7}$$

Closer examination of equation (7.6) discloses that for fixed values of K_D, R and $\underline{Z}$, the mixed voltage $\underline{U}_M$ only depends on $\underline{I}_1$ and $\underline{I}_2$. Its maximum value occurs for a ground fault, since $\underline{I}_1 = \underline{I}_2$, and its minimum value for a three-phase fault, since $\underline{I}_2 = 0$.

The considerations up to the present have concentrated on the problem of comparing phase-angles in a phase comparison scheme. It can be seen from Fig. 7.6, however, that the tripping signal has to be enabled by a starting function. This can be performed by the same kind of starting units as are used for distance protection (see Section 6.3.3), i.e. overcurrent, underimpedance or the phase-angle between current and voltage (phase-angle controlled starting) [7.13]. Some starting units in use are based on the detection of the symmetrical components of the fault current [7.10]. These detect three-phase faults by responding to the positive-sequence component, whereas all other kinds of faults are characterized by the presence of a negative-sequence component. For lines longer than 200 km, phase-angle sensitive underimpedance units with lenticular operating characteristics in the R/X plane are used instead of positive-sequence current units.

Some phase comparison schemes operate without starting units [7.1].

7.1.2.4 Distance protection

The fundamentals of distance protection were dealt with exhaustively in Section 6.3. For this reason, only those additional aspects of particular consequence for line protection will be presented here. These concern mainly the possible applications, influencing factors and limitations.

Multi-zone distance relays are used for detecting faults on transmission lines, typically with three zones measuring in the forwards and one in the reverse direction. The reaches of the three forwards measuring zones are chosen such that maximum discrimination and speed is achieved when faults occur. Figure 6.14 shows the principle of the time-step characteristic for grading the operation of distance relays in a radial power system and Fig. 6.16 for ring and meshed power systems. The reach of the first zone is normally set to between 80 and 90 % of the line impedance (line length), i.e.

$$Z_I = (0.8 \ .... \ 0.9)Z_L$$

The reach can, however, be extended to 130 % of the line impedance Z_L (overreaching), if the relay is part of an auto-reclosure scheme (see Section 7.1.4).

The reach of the second zone Z_{II} must on no account extend into the second zone of the next distance relay down the line. For example, the setting Z_{1II} of relay RZ1 in station A (Fig. 6.14) must fulfil the following condition

$$\underline{Z}_{1II} = 0.8(\underline{Z}_{AB} + \underline{Z}_{2I}) \tag{7.8}$$

where
Z_{AB} = positive-sequence impedance of line L1 (between stations A and B)
Z_{2I} = reach of the first zone of relay RZ2 in station B

Similarly, the reach of the third zone Z_{III} must not extend into the second zone of the next distance relay down the line. This means in the case of the distance relay RZ1 that

$$\underline{Z}_{1III} = 0.8[\underline{Z}_{AB} + 0.8(\underline{Z}_{BC} + \underline{Z}_{3I})] \tag{7.9}$$

where
Z_{BC} = positive-sequence impedance of line L2 (between stations B and C)
Z_{3I} = reach of the first zone of relay RZ3 in station C

Since the above impedance settings are made on the secondary side of the c.t's and v.t's, the primary system values must be multiplying by K_{Ni}/K_{Nu} to transform them to secondary values, K_{Ni} being the nominal c.t. ratio and K_{Nu} the nominal v.t. ratio.

The reaches for the second and third zones determined by applying equations (7.8) and (7.9) have to be reduced if there is an intermediate infeed in one of the neighbouring stations as shown in Fig. 7.9. In this case, the measurement is falsified, because relay RZ1 in station A only measures the current $\underline{I}'$ while the fault current flowing in line L2 is the sum of the currents $\underline{I}' + \underline{I}'' = \underline{I}_{BF}$. The relationship for the positive-sequence impedance measured by RZ1 up to the fault location has now changed to

$$\underline{Z}_M = \frac{\underline{I}'\underline{Z}_{AB} + (\underline{I}'+\underline{I}'')\underline{Z}_{BF}}{\underline{I}'} = \underline{Z}_{AB} + \left(1+\frac{\underline{I}''}{\underline{I}'}\right)\underline{Z}_{BF}$$

or

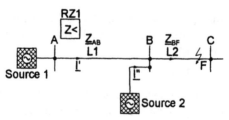

Fig. 7.9:   Influence of the intermediate infeed at station B on the distance measured by
             relay RZ1

$$\underline{Z}_M = \underbrace{\underline{Z}_{AB} + \underline{Z}_{BF}}_{\underline{Z}_{AF}} + \underbrace{\frac{\underline{Z}_{BF} I'}{I}}_{\Delta \underline{Z}_M} \qquad (7.10)$$

It can be seen that the positive-sequence impedance of the line between station A and the fault location F is increased by ΔZ_M which explains the reduction in reach of relay RZ1 mentioned above. It also explains why in today's meshed power systems with many infeeds it is practically impossible to make full use of a relay with more than three distance zones. A possibility of largely overcoming the intermediate infeed problem is to use adaptive protection systems.

In order to measure the distance of a ground fault between relay and fault locations correctly, the current variable used for measurement must be increased by a portion of the neutral current determined by the correction factor k_0:

$$\underline{k}_0 = \frac{1}{3}\left(\frac{\underline{Z}_0}{\underline{Z}_1} - 1\right) \qquad (7.11)$$

where
Z_1 = positive-sequence impedance of the protected line
Z_0 = zero-sequence impedance of the protected line

The principle of zero-sequence compensation is illustrated in Fig. 7.10 [6.7], from which it can be seen that the current I_M used for measurement includes the phase current I_{ph} and the neutral current I_Σ multiplied by the factor k_0. Thus both the amplitude and phase-angle are compensated in this case, whereas in others only the amplitude is compensated on the assumption that $Z_1 \cong X_1$ and $Z_0 \cong X_0$ and therefore $k_0 = (X_0 - X_1)/3X_1$.

In spite of compensating the current variable, correct measurement of ground faults by distance relays can still be in jeopardy on a double-circuit line. The difficulty is the mutual impedance between the neutral current circuits of the two parallel lines and the influence on the zero-sequence impedance of the ground conductivity, the spacing between phase and ground conductors, the penetration

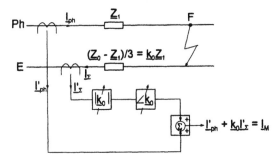

Fig. 7.10:   Principle of zero-sequence compensation for ground faults [6.7]

in the ground of the ground fault current and the radius or equivalent radius of the ground conductor. The distance of a fault in the case of several parallel current circuits can be increased or decreased by as much as 50 %. The first possibility delays the operation of the relay and the second causes it to mal-operate (due to overreach) [7.15].

The following are some of the methods employed for compensating the measurement of ground faults on double-circuit lines [7.16]:
-   inclusion of the neutral current of the other circuit, but this will cause indiscriminate operation of the protection, if the parallel circuit is switched off at one end
-   additional compensation switched in by the neutral current of the other circuit
-   correction by switching the value of the zero-sequence compensation factor
-   deliberate setting of k_0 to a value other than the one calculated for a single circuit
-   exchanging signals between the protection equipment at the two ends of the line

7.1.2.5 Transfer tripping schemes with distance relays

The drawback of the straightforward distance protection scheme described in the previous Section is that, because of the first zone setting of 85 to 90 % of the line impedance, faults in the last 10 to 15 % are only detected in the second zone and therefore are tripped after a delay (2nd. time step). By installing a communications channel between the distance relays at the two ends of a line, it is possible to trip faults along the whole length of the line in the shortest possible time (1st. time step). Depending on the preferred scheme and the application, the binary signal transferred from one relay to the other can either enable or block tripping by the local relay or switch its measuring reach. Of the many alternative schemes [7.1, 7.17, 7.18] the most important are

1. permissive transfer tripping
2. acceleration (zone extension)
3. directional comparison of distance measurements

In a *permissive underreaching transfer tripping scheme*, the first zone of both relays is set to the normal reach of 85 to 90 % of the line impedance. The local relay then trips its circuit-breaker for a fault in the nearest 10 to 15 % of the line in the minimum possible time (1st. time step) and simultaneously transmits a signal to the relay at the opposite end of the line. Providing at least the starters of that relay have picked up, i.e. the relay has detected the existence of a fault, the tripping signal received is "permitted" to trip the circuit-breaker.

Figure 7.11 shows the basic principle of a permissive transfer tripping scheme. It was assumed that the fault F occurred close to station A (Fig. 7.11a) so that it is detected in the first zone by relay RZ1 and in the second zone by relay RZ2. Relay RZ1 trips the local circuit-breaker CB1 without any intentional delay and transmits a transfer tripping signal via the transmitter T, receiver R and the AND gate AG2 to the circuit-breaker CB2 in station B. CB2 can only be tripped, however, if the starter Z_A of relay RZ2 has picked up. Since a certain time is required to transfer the intertripping signal from one station to the other (transmission time), tripping of the circuit-breaker CB2 via the communications channel is 10 to 40 ms slower than direct tripping by zone 1 [7.17].

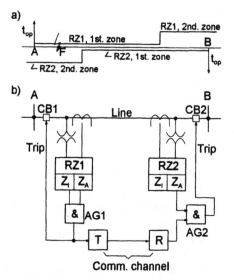

Fig. 7.11:   Principle of a permissive transfer tripping scheme [7.1, 7.17]
a) time step settings,   b) signalling connections,
Z_I = 1st. zone measuring unit, Z_A = starting unit

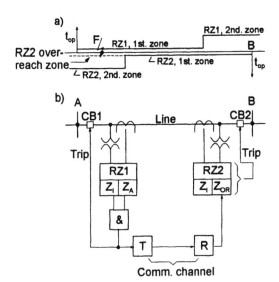

Fig. 7.12:   Principle of an acceleration scheme [7.19]
a) time step settings,   b) signalling connections,
Z_{OR} = overreaching zone

The first zone setting in an *acceleration scheme* is once again 85 to 90 % of the line impedance for both relays. In contrast to the above scheme, the inter-tripping signal in this case goes neither directly nor indirectly to the circuit-breaker in the remote station, but is used instead to extend the reach of the first zone to about 130 % of the line impedance (overreaching). Figure 7.12 illustrates this technique for a fault F close to station A (Fig. 7.12a). As before, relay RZ1 in station A detects the fault in its first zone and trips the local circuit-breaker CB1 without any intentional delay and transmits at the same time an intertripping signal to the distance relay RZ2. Upon receiving the signal, RZ2 extends the reach of its first zone to 130 % in the direction of station A (dashed line) and is then also able to measure the fault F and trip the circuit-breaker CB2 in station B. The tripping time in station B is twice the measuring time plus the transmission time. Compared with a permissive scheme, an acceleration scheme is slower by a relay measuring time (distance relay RZ2), but is still considerably lower than the normal tripping time of zone 2. An advantage of this scheme is its relatively high insensitivity to interference on the communications channel.

There are two kinds of *directional comparison schemes* using distance relays

- exchange of blocking signals
- exchange of permissive signals

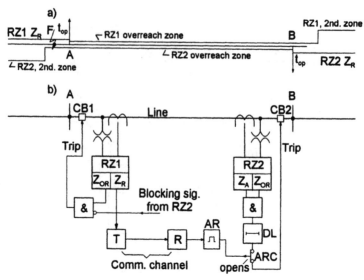

Fig. 7.13:   Principle of a directional comparison scheme with transmission of blocking
              signals
              a) zone settings,   b) signal transmission

All directional comparison schemes are unit protections (see Page 4). Where they are achieved by adapting the first zones of the distance relays, back-up protection is provided by the higher distance zones, for which otherwise separate relays would have to be installed.

The principle of a *blocking scheme* is illustrated in Fig. 7.13. The first zone in this case is always set to overreach the opposite station, but to prevent indiscriminate tripping for a fault on the next section of line (fault F in Fig. 7.13), tripping is inhibited in station B. The corresponding blocking criterion is provided by a reverse-looking underimpedance stage of the protection RZ1 which picks up for the fault F. A signal is then transmitted via the transmitter T and the communications channel CC to the receiver R where it excites and auxiliary relay AR so that its contact ARC interrupts the tripping circuit of the distance relay RZ2. The tripping relay for the circuit-breaker CB2 is thus blocked. Tripping of CB1 in station A does not take place either, because the overreaching stage of RZ1 does not pick up for the reverse fault. The advantage of this scheme is that the PLC blocking signals do not have to be transmitted through the fault on the line.

An alternative blocking scheme is to set the first zones of both relays to the normal setting of 85 % of the line and to switch them automatically to 130 % of the line length after a short delay, providing a blocking signal is not received in the meantime. This enables all faults along the full length of the line to be tripped in a time which is appreciably shorter than the normal operating time of the second zone.

In a *permissive overreaching transfer tripping* scheme, both distance relays are set to 130 % of the length of the line, but tripping can only take place if the relay in the opposite station also detects a fault in its first zone, i.e. both relays must detect a fault simultaneously in the forwards direction which can only occur for a fault on the protected line between the stations. One condition is, of course, that sufficient fault current must flow from both ends of the line. Since this scheme does not function at all should the communications channel fail or the infeed be too low, tripping must be enabled independently of whether a signal is received or not after a given time. The total tripping time is the same for faults along the whole length of the line and the scheme is suitable for transmission lines with series compensation [7.19].

7.1.2.6 Directional comparison protection

Apart from the directional comparison schemes based on a distance measurement described in the preceding Section, there are schemes which do not use distance relays. Of these, the most common are directional overcurrent phase fault protection, directional ground fault protection and the directional comparison of transient signals.

In the case of directional overcurrent phase fault protection (see Section 7.1.2.1), the directions of energy flow determined at the two ends of a line are compared and tripping of their respective circuit-breakers is enabled providing both flow from their busbars towards the line. Schemes of this kind detect faults along the whole length of a line quickly and absolutely selectively.

A directional ground fault scheme functions in a similar manner. It is used mainly where high-resistance ground faults of some tens to hundreds of ohms have to be detected. Such faults cannot be detected by a distance relay [7.20] and that is why a separate ground fault scheme is added. Distance and ground fault schemes can either use a common or separate communications channels.

The primary system quantity used for detecting the direction of ground fault energy is the phase relationship of the neutral current (zero-sequence component). The reference signal is the neutral voltage obtained from a broken delta v.t. circuit. The layout of the scheme is similar to that shown in Fig. 6.9 with the exception that the characteristic angle in solidly grounded systems is in the range 75° to 90° and in resistance grounded systems it is 0°. The circuit of a ground fault scheme for one line terminal in shown in Fig. 7.14. In the example of Fig. 7.14 the neutral current $3I_0$ is obtained from a Holmgren circuit, but core-balance c.t.'s are frequently used because of the greater sensitivity they permit.

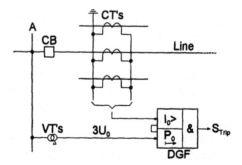

Fig. 7.14:   C.t. circuit for a directional ground fault scheme

A directional comparison scheme of this kind can operate by exchanging either blocking or permissive signals. An example of a permissive scheme is shown in Fig. 7.15. It can be seen from the diagram that undelayed tripping of the local circuit-breaker can only take place, if the local directional ground fault relay GFR generates an output signal S_{Trip} at the same time as an enabling signal ES is being applied to the AND gate AG. However, to ensure that the protection can trip in the event of a communications channel failure, provision is made for the relay to trip its local circuit-breaker on its own without any directional comparison between the relays DGF1 and DGF2 after a delay determined by the timer DL.

The disadvantage of this scheme is that it cannot clear faults on the protected line at its normal high speed, if a circuit-breaker is open at one end. What is referred to as an echo circuit as shown in a simplified form for station B in

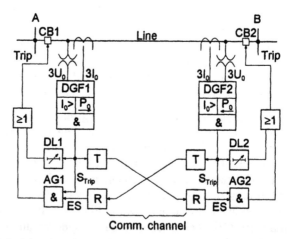

Fig. 7.15:   Principle of a directional comparison scheme with exchange of permissive signals between the directional ground fault relays at the two ends of a line [7.19]

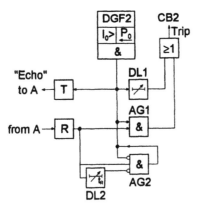

Fig. 7.16: A directional comparison scheme with echo circuit in station B (see Fig.
7.15)

Fig. 7.16 is added to overcome this disadvantage. A comparison of this circuit
with that of Fig. 7.15 discloses that the term echo comes from the fact that the
enabling signal transmitted from station A is looped back to station A via the receiver R, the AND gate AG2 and the transmitter T in station B. The signal arriving back at A simulates the enabling signal needed from station B to permit tripping of the circuit-breaker CB1. The arrangement of Fig. 16 assumes that the
circuit-breaker CB2 is open, i.e. it is impossible for the protection DGF2 to pick
up. The duration of the echo signal is limited to t_R by the timer DL2.

Up to the present it has been assumed that the measuring circuits of the directional comparison scheme have only had to process steady-state sinusoidal
input variables. In recent years, protection principles have been developed, however, which are capable of determining direction from the evaluation of non-
steady-state variables [7.21, 7.22, 7.23]. These rely on the fact that the currents
and voltages at the beginning of a transmission line are subject to step changes
when a fault occurs as defined by the following equations

$$\Delta i(t) = i(t) - i_p(t) \qquad\qquad (7.12)$$

$$\Delta u(t) = u(t) - u_p(t) \qquad\qquad (7.13)$$

where

$i(t), u(t)$ = input variables in relation to time after the incidence of the fault
$i_p(t), u_p(t)$ = input variables in relation to time before the incidence of the
fault
$\Delta i(t), \Delta u(t)$ = variation of the input variables in relation to time as a consequence of the fault

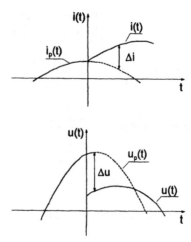

Fig. 7.17: Excursions of current and voltage at the relay location before and after the incidence of a fault (fault incidence at t = 0) [7.22]

The variation of current and voltage at the ends of a line before and after the incidence of a fault is illustrated in Fig. 7.17.

If the current signal is made to flow through a replica impedance Z_M of the protected line (see Section 6.3), a voltage difference $\Delta u_M(t)$ is obtained which can be used in conjunction with $\Delta u(t)$ to determine fault direction. The excursion of the input variables in relation to time is easily represented in the delta plane with the values for Δu along the abscissa and the values for Δu_M along the ordinate. Figure 7.18 shows the trajectories in the delta plane for a fault in the forwards direction. In this case, the trajectory moves from the second to the fourth quadrant and then back to the second. For a fault in the reverse direction, the

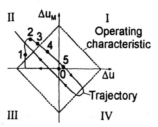

Fig. 7.18: Trajectories of the variables in the delta plane for a fault in the forwards direction [7.22]
1, 2, ... 5 = time in ms after the incidence of the fault

movement is from the first to the third. The sign of the voltage u(t) at the instant of the fault determines in which quadrant the trajectory first appears. By selecting a suitable operating characteristic (e.g. the dashed line in Fig. 7.18), all faults in the forwards direction can be reliably detected. Tripping is enabled if the characteristic is transgressed in the second quadrant and it can be seen that this principle enables faults to be detected in an extremely short time (1 to 3 ms) after the incidence of the fault. A further characteristic is used to detect high fault levels close to the relay location, but in this case tripping is independent of the directional unit in the station at the opposite end of the line. According to the configuration of the power system, this method enables faults at locations up to 40 % of the line length to be tripped without signals being exchanged between the protection devices at the ends of the line.

An additional tripping criterion was included to ensure phase-selective operation even in the presence of HF interference in the input signals [7.22]. It is based on supervising the following integral function in relation to time

$$F(\tau) = \int_0^\tau \Delta u(t) \cdot \Delta U_M(t) \, dt \tag{7.14}$$

This function is always located in the negative half of the plane for faults in the forwards direction and in the positive half for faults in the reverse direction. Figure 7.19 shows the excursion of the integral function $F(\tau)$ for a fault in the forwards direction. The fault direction can be determined quickly and reliably by appropriately setting the pick-up levels L and M.

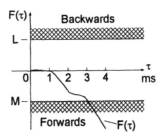

Fig. 7.19: Excursion of the integral function $F(\tau)$ for a fault in the forwards direction [7.22]
L, M = pick-up levels for determining direction

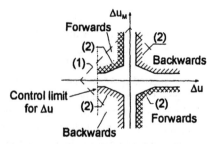

Fig. 7.20:   Operating characteristic of a high-speed directional comparison protection
             which measures the differential of the input variables [7.22]
             (1) Operating characteristic not relying on the directional measurement in
                  the opposite station
             (2) Operating characteristic relying on the directional measurement in the
                  opposite station

Figure 7.20 shows the operating characteristic of a practical high-speed directional comparison scheme and Fig. 7.21 its block diagram [7.22]. It follows from Fig. 7.20 that the operating characteristic is made up of two components, i.e. characteristic (1) from Fig. 7.18 for tripping irrespective of the directional measurement in the opposite station and characteristic (2) for directionally dependent operation.

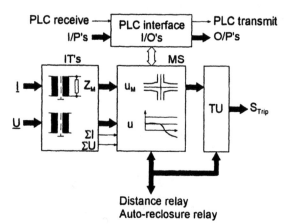

Fig. 7.21:   Block diagram of a high-speed directional comparison protection which
             measures the differential of the input variables [7.22]
             IT's = input transformers,   MS   = measuring system,
             TU   = tripping unit

7.1.2.7 Back-up protection for detecting phase faults

As was explained in Section 1.2, at least two redundant protection systems are essential, because the failure of a protection system can never be entirely excluded. This is especially true for phase fault protection on important transmission lines which is almost invariably always duplicated. It is usual to choose different measuring principles which monitor different system criteria for the duplicate phase or ground fault protection schemes. A typical technique on EHV transmission lines is to combine a conventional distance protection scheme with a phase comparison scheme, both of which independently detect and trip the majority of faults along the line with the minimum delay and yet complement each other's performance in certain kinds of fault situations [7.26].

An alternative combination of different protection schemes comprising a distance scheme as the main protection and a directional ground fault scheme either with or without a communications channel as back-up is often applied on HV transmission lines.

Apart from the "local" back-up protection schemes described above, i.e. the back-up protection is installed at the same location as the main protection and protects the same unit, at least one of the line protection schemes is designed to serve as back-up for the protection devices on the adjacent item of primary system plant. This is a feature of both a graded distance protection scheme and graded time-overcurrent protection which have already been explained in detail.

7.1.3 Auto-reclosure

7.1.3.1 Purpose and types of auto-reclosure

Statistically, the majority of faults which befall an overhead line transmission system are of a transient nature and disappear if the supply of energy to the fault location is briefly interrupted. The automatic reclosure of the circuit-breakers at the ends of a line is initiated by the protection devices which detected the fault and tripped the circuit-breakers in the first place. The line is reenergized after a given period call the dead time which is mainly determined by the time required for the air surrounding the arc to deionize after the arc has extinguished [7.27...7.31]. The device which performs this function is called an auto-reclosure relay.

The purpose of auto-reclosure is thus to reenergize a line as quickly as possible after a fault caused it to be tripped and thereby reestablish the normal state of the power system and the supply to isolated areas. It is assumed that this will normally be successful and only in seldom cases will it be necessary for the protection to trip the line again, because the fault persists. Accordingly the two possibilities are referred to as **successful** auto-reclosure and **unsuccessful** auto-reclosure. In the former case, the fault disappears during the dead time (arc extinguished and the voltage withstand re-established at the fault location), while in the latter case, the circuit-breakers close onto an existing fault and the protection trips the line again. The two trips in quick succession are interpreted by the auto-reclosure relay as a permanent fault and no reclosure is performed after the second trip, i.e. the circuit-breaker is "locked out". The sequence described here is that of a "single-shot" auto-reclosure scheme; in medium voltage systems, two-shot schemes are sometimes employed. Figure 7.22 shows the operating sequence of the circuit-breaker (line 1), the protection (line 2) and the auto-reclosure relay (line 3) for a successful single-shot auto-reclosure attempt. Line 4 of the diagram gives an impression of what happens to the transmission line current before, during and after auto-reclosure.

A further distinction which is made between auto-reclosure schemes depends on whether one or all three poles of the circuit-breakers are tripped and reclosed. Single-phase reclosure is only applied in solidly grounded power systems and usually only at EHV level, but occasionally it is encountered on HV transmission lines. Only three-phase auto-reclosure is applied on ungrounded systems or systems with Petersen coils at the MV level. Three-phase reclosure has to be used in HV systems, if the circuit-breakers are not suitable for single-phase reclosure, i.e. the circuit-breaker poles are not equipped with individual mechanisms.

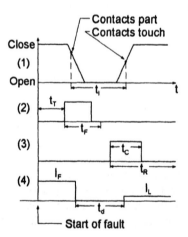

Fig. 7.22:   Operating sequence of a successful auto-reclosure attempt [7.28]
t_i = current interruption time, t_T = tripping signal, t_d = auto-reclosure dead time, t_C = auto-reclosure closing impulse, t_R = auto-reclosure reclaim time, t_F = maximum fault duration for permitting auto-reclosure, I_F = fault current, I_L = load current

Three-phase auto-reclosure is performed both with and without checking the synchronism of the systems on the two sides of the interruption depending on the duration of the dead time, which in the case of slow reclosure is in the order of seconds. The dead time is defined as the time from the instant the current is interrupted when tripping the circuit-breaker until it flows again when reclosing the circuit-breaker (line 4 in Fig. 7.22). Typical dead times in practice range in the case of single-phase auto-reclosure from 0.4 to 1.2 s and for three-phase reclosure without a synchrocheck feature from 0.3 to 0.5 s, respectively with a synchrocheck feature from 1 to 5 s [7.27].

7.1.3.2 Coordination between auto-reclosure and line protection functions

Auto-reclosure is applied in conjunction with different kinds of protection relays. The following discussion, however, only considers the application of auto-reclosure in conjunction with overcurrent relays and with distance relays.

As was explained in Section 7.1.2.1, radial lines in MV systems are protected by non-directional overcurrent relays. In a time-overcurrent protection scheme with auto-reclosure (Fig. 7.23), the circuit-breaker is tripped undelayed upon the detection of a fault, i.e. without waiting for the grading time to expire. The circuit-breaker is closed again at the end of the dead time t_d. Providing the reclosure attempt is successful, the normal load current I_L flows after reclosure (Fig. 7.23d); if the fault still persists after reclosure, the circuit-breaker CB1 is tripped again by the overcurrent relay, but this time after the set grading time t_1, i.e. the instantaneous trip function t_{T0} is blocked (Fig. 7.23e). Figure 7.24 is a simplified representation of a single-shot three-phase auto-reclosure relay in conjunction with the time-overcurrent relay RI operating according to the timing sequence of Fig. 7.23.

Auto-reclosure in conjunction with a distance relay involves switching the reach of the first zone. Two versions of this principle are in use, one in which the distance relay normally overreaches and the reach is shortened and one in which the distance relay normally underreaches and the reach is extended [7.27].

In a scheme which **shortens the reach** (Fig. 7.25b), the first zone is set to 130 % of the protected line so that all faults along the entire length of the line are initially tripped in the shortest possible time. After the trip, the auto-reclosure relay automatically switches the reach of the distance relay to 80 to 90 % of the length of the line (normal first zone reach) and then recloses the circuit-breaker. The corresponding timing sequence for a radial line is given in Fig. 7.25b. Where fault energy can be supplied from both ends of the line, there is a similar arrangement in the opposite station. This scheme accepts that faults in the first 20 % of a line are initially tripped by two stations in series, but since the relay on

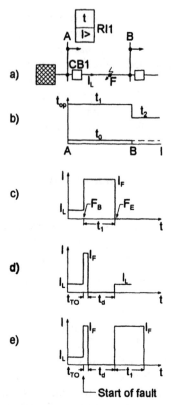

Fig. 7.23:   Timing sequence for tripping a fault on a radial line without auto-reclosure
c) and with auto-reclosure d) and e)
a) radial line,   b) overcurrent relay operating times,   c) line current without reclosure,   d) line current with successful auto-reclosure,   e) line current with unsuccessful auto-reclosure, t_1 = grading time set on RI1, t_0 =
time without intentional delay, t_{TO} = fault tripping time without intentional
delay, t_d = auto-reclosure dead time

the adjacent section of line no longer sees the fault in the first zone after reclosure and therefore its reclosure attempt will always be successful, this is not an
operational disadvantage.

In a scheme which **extends the reach** (Fig. 7.25c), the first zone is normally
set to 80 to 90 % of the protected line. If the fault is beyond the next station, it
will be tripped in the shortest possible first zone time by the relay there and the
relay on the upstream section of line which also picked up resets without tripping. It does not reset, however, for a fault in the last 20 % of the line before
station B and a short time after having picked up its reach is extended to 130 %

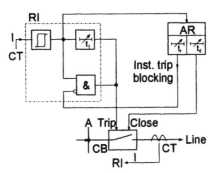

Fig. 7.24:   Simplified block diagram showing the interconnections between time-
             overcurrent and auto-reclosure relays

of the line length (overreach zone) to enable it to detect the fault and trip in the
first zone. The auto-reclosure relay switches the reach back to 80 to 90 % of the
line length at the end of the tripping signal.

Single and three-phase auto-reclosure are possible with both these techniques. Auto-reclosure relays are available which can be configured for operation
in both kinds of schemes or on an "either/or" basis. The latter arrangement facili-

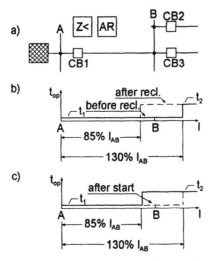

Fig. 7.25:   Timing sequence of distance and auto-reclosure relays on a radial line
             a) radial system,   b) reduction of reach after reclosure,   c) extension of
             reach after reclosure

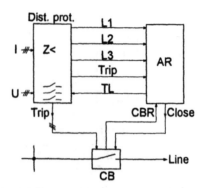

Fig. 7.26:   Flow of signals between distance relay, auto-reclosure relay and circuit-breaker for single-phase reclosure [7.29]
L1, L2, L3 = phase detector starting signals,   TL = tripping logic control for single or three-phase tripping,   CBR = circuit-breaker ready and position signals

tates single or three-phase tripping on the same line according to the kind of fault, i.e. a single-phase reclosure is performed for ground faults and three-phase reclosure for all combinations of phase faults. Figure 7.26 shows the flow of signals for a single-phase reclosure [7.29].

It should be noted that only single-phase reclosure is permitted on HV and EHV lines close to large power plants; the circuit-breaker must be locked out for all phase faults. The reason for this is the high torsional stress on the shaft train

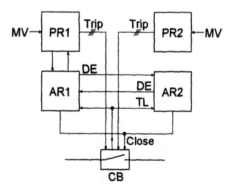

Fig. 7.27:   Flow of signals with first and second main protections and two auto-reclosure relays [7.29]
MV = measured variable,   DE = data exchange

of a large turbo-alternator set exerted by recurring phase fault currents close to the power plant [7.32].

Apart from the time-overcurrent and distance schemes with auto-reclosure described above, auto-reclosure is also used in conjunction with other protection measuring principles such as signal comparison, differential and phase comparison schemes. Since for reasons of reliability and availability, HV and EHV transmission lines are invariably equipped with first and second main protections (see Section 7.1.2.7), correspondingly coordinated auto-reclosure schemes are also necessary. One of these employs two independent sets of protection and auto-reclosure relays as shown in Fig. 7.27. The arrangement assumes that it is possible for the protection devices to respond differently to different faults and therefore that certain faults may only be detected by one of the schemes. This necessitates exchanging signals between the auto-reclosure relays to ensure that they respond in a coordinated manner [7.29]. The corresponding logic

- blocks the second auto-reclosure relay once the dead time on the first auto-reclosure relay has started.
- makes the reclaim time of the first auto-reclosure relay effective for the second auto-reclosure relay.

The above precautions avoid the risk of a single-phase close command being generated after a three-phase reclosure has already taken place.

Where there is only one main protection and the second protection scheme operates in a delayed back-up role, auto-reclosure is only used in conjunction with the main protection.

7.1.4   Protection schemes for detecting ground faults

7.1.4.1 Signals for detecting ground faults

In Section 2.2 it was explained that ground faults in ungrounded systems and systems with Petersen coils are characterized by the occurrence of a steady-state neutral current and a neutral voltage and also by transient currents and voltages. These signals are thus available or can be derived in order to detect the existence of a ground fault and locate the line effected. Other signals also used for this purpose which will be gone into in more detail later are the harmonics caused by the ground fault on the power system or injected HF signals. Thus the following criteria are used for detecting and locating ground faults [7.33]:

- steady-state ground fault current
- steady-state ground fault power
- transient ground fault current
- power system harmonics or injected HF signals

The following protection devices are applicable for detecting these signals [7.34 ... 7.37]:

- ground fault overcurrent
- directional ground fault
- fleeting ground fault
- directional harmonic ground fault protection

The various aspect and possibilities in ungrounded systems and systems with Petersen coils are the subject of the next Section.

7.1.4.2 Ground fault overcurrent protection

Ground fault overcurrent protection is only applicable to ungrounded radial systems and only then providing the ground fault current is sufficiently higher than the relay setting. According to Section 2.2 (equation (2.14)), the ground fault current at the fault location in a given power system is dependent on the total ground capacitance of the electrically connected network and the power system voltage. Its value is always less than the maximum permissible load current of the protected line, because if it is not, Petersen coils are installed to improve system operating conditions.

Two alternative arrangements are used for measuring ground fault current, the Holmgren circuit (Fig. 7.28a) or a core-balance c.t. (Fig. 7.28b). The core-balance c.t. exhibits high sensitivity and low errors and therefore it is especially suitable for measuring low levels of ground fault current. Typical ratios are 60/1 A or 75/1 A and correspondingly the secondary currents generated by the lower primary fault currents are in the order of tens to hundreds of milliamperes. Special overcurrent relays have to be used to measure these low currents [7.34]. Tripping by the protection is usually delayed to grade with the other ground fault relays.

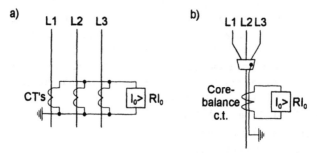

Fig. 7.28:   Current measurement for ground fault overcurrent protection
             a) Holmgren circuit,   b) core-balance c.t.

7.1.4.3 Directional ground fault protection

Directional ground fault schemes are used in ungrounded systems and systems with Petersen coils. In the first case, the direction of the capacitive component of the neutral power is monitored and in the second the real power component. In the event of a ground fault, the energy of the corresponding component of power flows towards the fault location. The relationships for an ungrounded system are shown in Fig. 7.29. It follows from the diagram that the directions of the currents at the beginning of the lines L1, L2 and L3 depend on the fault location. The direction of the energy in the faulted phase is always from station A to the fault location (I_{OF} for line L3 in Fig. 7.29a, respectively for L1 in Fig. 7.29b). If the neutral voltage U_0 (zero off-set voltage) is chosen as the reference voltage to determined the phase relationship of the current, the power direction is given by the phase-angle φ_0 between U_0 and I_0. The variable I_0 can be measured as in the case of ground fault overcurrent protection by means of a Holmgren circuit or a core-balance c.t. and U_0 from the broken delta connection of the secondaries of three v.t's (see Table 4.3). The c.t. and v.t. connections of a directional ground fault scheme can be seen from Fig. 7.30. The solid-state protection device DGF corresponds basically to Fig. 6.9a (Section 6.1.4) with the exception that the input variables are U_0 and I_0.

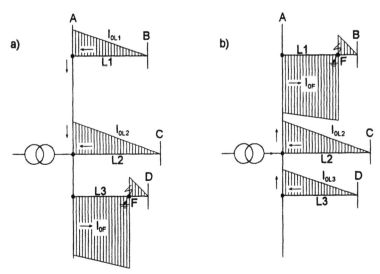

Fig. 7.29:   Distribution and direction of the ground fault currents in an ungrounded power system
a) ground fault located on line L3
b) ground fault located on line L1

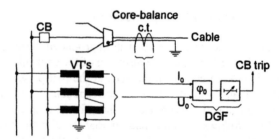

Fig. 7.30:   C.t. and v.t. connections of a directional ground fault protection DGF

For reasons of cost, a single directional ground fault relay is frequently used to monitor several feeders in the same station. Using a single relay requires that the secondary neutral currents of the individual feeders be connected to the relay in sequence either automatically or manually. The principle is given in Fig. 7.31. During normal system operation, the secondary windings of the core-balance c.t's are short-circuited by the buttons S1, S2 and S3. A ground fault on one of the feeders triggers the automatic scanning of the feeders by operating the buttons in sequence until the feeder with the fault is located. This arrangement is mainly used to signal the existence and location of a ground fault, but can also be used to selectively isolate the faulted line.

As mentioned previously, the real power component of the neutral current measured by the relay is used to locate ground faults in systems with Petersen

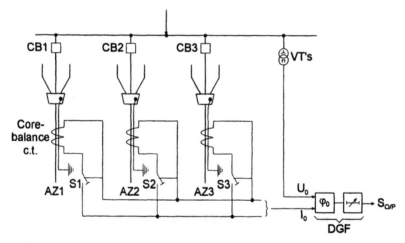

Fig. 7.31:   Principle of using a common directional ground fault relay for several feeders

coils. Its value is determined principally by the resistances of the Petersen coil and the power system losses in parallel. At the fault location in an MV system, the residual current amounts to 5 to 8 % of the capacitive ground fault current without the Petersen coil, the lower limit being true for cable systems and the higher one for overhead line systems. The real power component in the neutral current of HV power systems is even lower and in a small system may be too low for accurate fault location, especially as harmonics generated by the power system itself and the loads are superimposed on the 50/60 Hz component. Thus to ensure reliable location of ground faults in small systems as well, the real power component is artificially increased (up to 20 A) by connecting a grounding resistor in parallel to the Petersen coil for a brief period after the existence of a ground fault has been detected. This arrangement can be seen in Fig. 7.32. The grounding resistor R is connected across the Petersen coil PC by the switch S about 2 to 3 seconds after the fault occurs. The switch S is controlled by the neutral voltage relay RU_0 via the time delay DL1. The second time delay DL2 applies the neutral voltage U_0 to the protection relays DGF1 and DGF2 which obtain their neutral currents I_0 from the core-balance c.t's CBCT. The switch S remains closed a few seconds, sufficiently long for the protection relays to reliably determine the location of the fault.

The same principle can be applied in conjunction with the central fault detector unit of Fig. 7.31, providing the grounding resistor R is correspondingly rated for the longer time needed by the feeder scanning procedure.

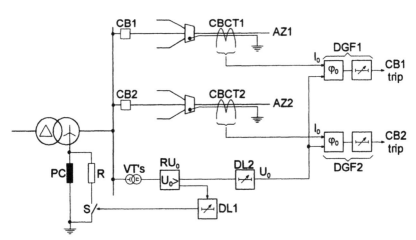

Fig. 7.32:   Artificially increasing the real power component of the neutral current to facilitate ground fault location.

7.1.4.4 Fleeting ground fault protection

The ground fault protection schemes discussed up to the present processed the steady-state system quantities to locate a fault and are therefore only able to detect permanent ground faults. By measuring the transient behaviour of the system variables initiated by a fault, the possibility exists of detecting a fault before it becomes a permanent one and also of locating fleeting ground faults. The principle is based on the fact that in both ungrounded systems and systems with Petersen coils, a ground fault always begins with a high current surge (damped ignition wave) which is several times higher than the steady-state ground fault current [7.35, 7.38]. The sign of the current surge in the first half-wave is used as the criterion for determining fault direction. The principle of the protection circuit used to detect a fleeting ground fault is shown in Fig. 7.33. The input variables are the neutral current I_0 and the off-set voltage U_0 measured between the star-point of the three condensers and the ground of the three v.t's. The transient neutral current produces a voltage drop across the inductance L which results in a proportional current through the interposing c.t's ICT1 and ICT2. The output signals of the interposing c.t's are resultant variables given by the following relationships

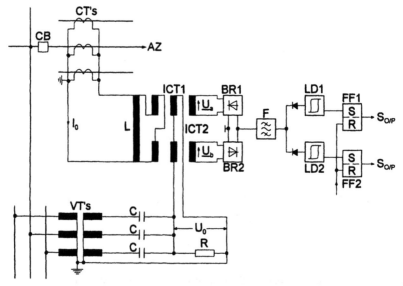

Fig. 7.33:   Principle of the protection circuit for detecting fleeting ground faults [7.35]

$$\underline{U}_a = \left(\underline{U}_0 - \underline{I}_0\right) \tag{7.15}$$

$$\underline{U}_b = \left(\underline{U}_0 + \underline{I}_0\right) \tag{7.16}$$

The values of the output voltages are compared by the bridge rectifiers BR1 and BR2 and then filtered in F before being applied to the level detectors LD1 and LD2. The ground fault is on the protected section of line if

$$\left|\underline{U}_a\right| > \left|\underline{U}_b\right| \tag{7.17}$$

The outputs of the two level detectors LD1 and LD2 are interlocked such that when one of them picks up it blocks the other. The information as to which of the level detectors picked up is stored in the respective bistable flip-flop FF1 or FF2. Since the fleeting ground fault protection scheme only evaluates the first current wave, it cannot be used as a central ground fault scanning relay. It is, however, frequently used in combination with a normal ground fault protection in systems with Petersen coils.

7.1.4.5 Harmonic directional ground fault protection

This scheme uses the steady-state harmonics in neutral current and voltage (frequently the fifth) to discriminatively detect ground faults.

The harmonics are generated by the non-linearities of power transformers. Since a Petersen coil only compensates the fundamental of the capacitive ground fault current and has no influence on the harmonics, the fault direction is given by the direction of reactive power flow of the harmonics [7.37] regardless of whether the system is ungrounded or grounded via a Petersen coil. The ground fault is in the direction of the line, if the reactive power of the harmonics during a ground fault is measured to be inductive at the relay location, and in the direction of the busbars, if it is capacitive.

This kind of scheme can be accomplished by a circuit similar to the ground fault scheme of Fig. 7.30 with the exception that filters, which reject the fundamental and the third harmonic and amplify the fifth harmonic, are inserted in the inputs of the variables I_0 and U_0 before they are applied to the phase comparator.

In power systems with a fifth harmonic content which varies with fluctuations of load during the course of the day, a signal with a frequency up to 500 Hz is injected, the level of which is not subject to power system load or configuration.

7.2 Protection of multi-winding and auto-transformers

7.2.1 Methods of protection

The methods of protection used to protect multi-winding and auto-transformers depend on the types of faults, the system disturbances which can occur and the rated power of the protected unit. It is obvious that much more protection will be installed to protect a costly high-power transformer than a transformer of only a few hundred kVA or MVA. This is also reflected in the choice of whether or not redundant protection systems are installed.

With regard to the types of faults which can occur either inside the transformer or on the external power system, the protection schemes can be divided into two main groups, i.e. protection schemes for detecting

- internal faults
- external faults

The internal and external faults which represent a hazard to all kinds of power transformers are listed in Table 7.2.

The kinds of protection used to detect the above faults are given in Table 7.3, the ones actually chosen depending on the size of the transformer concerned, e.g. time-overcurrent protection for the back-up protection of a small transformer and distance protection for the back-up protection of a large one.

The following Sections describe the most important features of the various schemes. The Buchholz protection is also gone into in detail, although its operating principle diverges considerably from the other protection schemes for processing electrical quantities described in this book. The overexcitation (overfluxing) protection, on the other hand, is dealt with in Section 8.2.4, because it is mainly used in practice on the step-up transformers of generators.

Table 7.2   Internal and external faults which threaten all kinds of power transformers

Internal faults	External faults
Short-circuit between windings Short-circuit between turns Ground faults Tap-changer failure Transformer tank oil leaks	Power system phase faults Power system ground faults Overload Overexcitation (overfluxing)

Table 7.3   The protection schemes used for power transformers in general

Kind of fault	Protection
Phase and ground faults Ground faults	Biased differential[1] Time-overcurrent[2] or [1] Distance[2] Ground fault
Short-circuit between turns Transformer tank oil leaks	Buchholz
Overload	Overload Thermal image of protected unit
Overexcitation	Overexcitation[3]

[1] main protection   [2] back-up protection   [3] only step-up transformers

7.2.2   Protection schemes for detecting phase and ground faults

7.2.2.1 Biased differential protection

The basic principle of differential protection was explained in Sections 3.2 and 6.2. Its use in connection with power transformers, however, was only touched upon in connection with inrush restraint. In the case of power transformers, there may also be a phase rotation between primary and secondary, e.g. group of connection Yd5 or Yd11, which has to be compensated before the currents measured in primary and secondary can be compared by the differential relay. Should the ratios of the c.t's on primary and secondary sides not correspond to the effective rated currents, the amplitudes must also be compensated so that difference between them at the relay becomes zero under normal load conditions. Finally, the zero-sequence component must be eliminated on windings with grounded star-points

Formerly, the compensation of group of connection and main c.t. ratios was performed outside the differential relay by appropriate connection of the main c.t. secondaries or of interposing c.t's inserted between the main c.t's and the differential relay DR (see Fig. 7.34). In the latter case, it can be seen from the diagram that the main c.t's are Y-connected while the three single-phase interposing c.t's are connected with the same group of connection as the power transformer. Typical interposing c.t's have several tappings for adjusting their ratio K_{ICT} to the

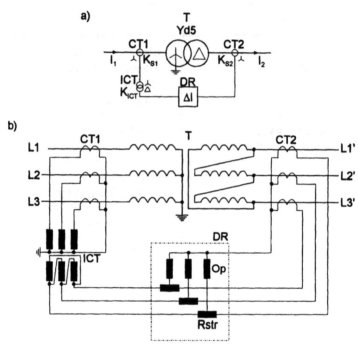

Fig. 7.34:   Phase and ratio compensation of the biased differential protection for a
            power transformer with the group of connection Yd5
            a) single line diagram   b) three-phase diagram

composite ratio of the power transformer K_T and the main c.t's K_{S1} and K_{S2}. The
following relationship for determining the ratio K_{ICT} is derived from Fig. 7.34a.

$$K_{ICT} = \frac{K_{S1} \cdot I_{2N}}{\sqrt{3} K_{S2} \cdot I_{1N}} = \frac{K_{S1}}{\sqrt{3} K_T \cdot K_{S2}} \qquad (7.18)$$

where
I_{1N}, I_{2N} = primary, respectively secondary rated current of the power trans-
                  former

    Interposing c.t's are also used in the case of Y-connected primaries and sec-
ondaries, e.g. group of connection Yy0, to compensate differences of ratio. Since
they are also connected in Yy0, the $\sqrt{3}$ term in equation (7.18) is omitted.

    Phase and ratio compensation is essential in the case of three-winding power
transformers, because there is always a phase-shift between the currents of the
three windings and the ratios of the three groups of main c.t's have to be adjusted

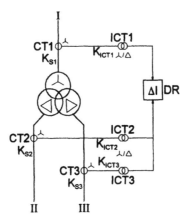

Fig. 7.35:   Phase and ratio compensation of a three-winding power transformer for a
biased differential scheme

to each other. Figure 7.35 shows in a simplified form the principle of compensation for a Ydd three-winding transformer with three groups of interposing c.t's
ICT1, ICT2 and ICT3.

The ratios of the interposing c.t's in the three legs n can be determined using
the following generally applicable formula [7.39].

$$K_{ICTn} = \left(\frac{S_N}{\sqrt{3} U_N \cdot K_{Sn}} \right) \frac{I}{m} \tag{7.19}$$

where
K_{Sn} = main c.t. ration in legs $n = 1, 2, 3$
m = factor depending on the interposing c.t. circuit:
          $m = I_{NP}/\sqrt{3}$ in the case of star/delta,
          $m = I_{NP}$ in the case of star/star, I_{NP} being the primary current of
          the interposing c.t's, i.e. secondary current of the main c.t's
S_N = rated transformer power
U_N = rated voltage of the respective transformer winding

Some years ago, the first transformer differential relays were developed
which permit the compensation for the group of connection of the power transformer and the ratio of the main c.t's to be made by adding and subtracting phase
currents electronically inside the relay [7.6, 7.40]. The principle of this method of
compensation for a Yd5 two-winding transformer is shown in Fig. 7.36. The
configuration of the phase currents I_{L1}, I_{L2} and I_{L3} is determined by the setting of
the selector switch SS1 and the currents I'_{L1}, I'_{L2} and I'_{L3} by the selector switch
SS2. In this case, phase compensation is performed on the Y side of the trans-

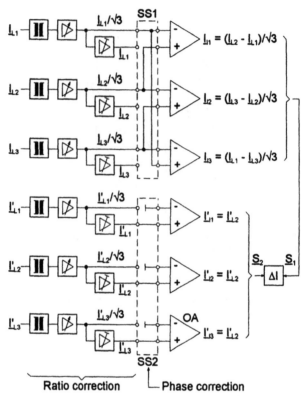

Fig. 7.36: Phase and ratio compensation in a biased differential relay for the group of connection Yd5 [7.6, 7.40]
OA = operational amplifier, SS1, SS2 = selector switches

former (currents $\underline{I}_{L1}$, $\underline{I}_{L2}$ and $\underline{I}_{L3}$). Since under normal load conditions, the currents compared, e.g. $\underline{I}_{L1}$ and $\underline{I}'_{L1}$, must have the same amplitude and phase-angle, the phase currents at the input of the relay are either applied directly to the selector switch or divided by $\sqrt{3}$ beforehand. As can be seen from Fig. 7.37, the ratio compensation in the example being considered is precisely correct.

In the case of a three-winding power transformer, the input variables have to be appropriately combined to obtain correct comparison. This follows from the circuit of Fig. 7.38 which assumes that under normal load conditions winding I is the infeed with $\underline{I}_1$ flowing towards the transformer. Neglecting the transformer losses, the sum of the currents $\underline{I}_2$ and $\underline{I}_3$ leaving windings II and III equals the primary current $\underline{I}_1$. Thus under normal load conditions the current in the differential (operating) leg can only be zero if

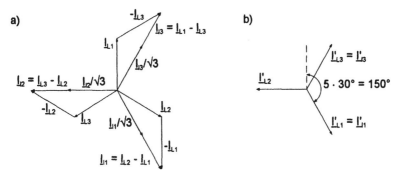

Fig. 7.37: Vector diagrams for the comparison of currents in a differential relay for
the group of connection Yd5
a) Variable S_1 star side of the transformer,   b) Variable S_2 delta side of
the transformer

$$I_a = I'_1 - \left(I'_2 + I'_3\right) \tag{7.20}$$

where
I'_1, I'_2, I'_3 = main c.t. secondary currents

In this example, the interposing c.t's which would otherwise be necessary for
phase and ratio compensation have been omitted for simplicity. It can be seen
that the protection is restrained in the event of a fault outside the zone of protection by the currents I_{rstr1} and I_{rstr2}. In normal operation

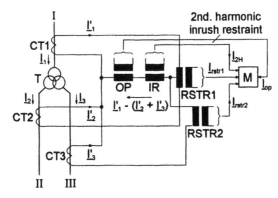

Fig. 7.38: Variables measured by a biased differential scheme for a three-winding
power transformer

$$I_{rstr1} = \underline{I}_1 + \underline{I}_2$$

$$I_{rstr2} = \underline{I}_3 \tag{7.21}$$

The two restraint currents can be arithmetically added to obtain the composite restraint current I_{rstr} which is described by the following relationship

$$I_{rstr} = \left(\left| \underline{I}_1 + \underline{I}_2 \right| + \left| \underline{I}_3 \right| \right) k_{rstr} \tag{7.22}$$

where
$k_{rstr} \leq 0.5$ = pick-up ratio of the differential protection

The following condition is always fulfilled and thus the discrimination of the protection assured for a through-fault

$$\left| I_{rstr} \right| > \left| I_a \right|$$

7.2.2.2 Time-overcurrent protection

Time-overcurrent protection is used as back-up protection for small power transformers with ratings in the range of a few tens of MVA or even as the main protection on very small transformers to detect both internal and external phase faults. While only one set of relays is installed on the primary side of two-winding transformers, each of the windings of a three-winding transformer has to be equipped. The pick-up setting depends on the rated current of the transformer, but the time delay is determined by the time grading of the power system described in Section 6.1.2 (Fig. 6.5b).

7.2.2.3 Distance protection

On transformers with higher ratings (> 100 MVA), the above function (i.e. back-up protection) is performed by a distance relay instead of overcurrent relays. Its use in the case of a step-up transformer in a power generation plant is dealt with in Section 8.2.3. This Section is only concerned with the application of a distance relay as the back-up protection of power system transformer.

Figure 7.39 shows the principle of a distance scheme for a two-winding transformer (or auto-transformer). Distance relays are installed on both sides of the transformer. This example assumes that three distance zones measure in the forwards direction and one in the reverse direction. The two distance relays DR1 and DR2 therefore both provide back-up protection for the differential protection, the line protection of the various feeders and the busbars.

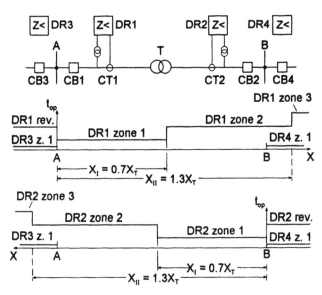

Fig. 7.39: Application of distance relays to detect internal and external phase faults on a two-winding power transformer
DR = distance relay, DR1 rev. = reverse zone of the distance relay
DR1, X_T = equivalent reactance of the transformer

7.2.2.4 Ground fault protection

The ground fault protection scheme depends on whether it is only required to detect ground faults on the transformer or should also protect the transformer from the effects of overcurrents due to ground faults on other items of plant.

If the zones of protection are confined to the transformer windings, the most common scheme for detecting ground faults is the one shown on both sides of the transformer in Fig. 7.40. It is called "restricted ground fault protection" and operates according to the high-impedance principle, i.e. the phase c.t's, and in the case of the star winding, a c.t. in the star-point (CT1 and CT2), respectively in the case of the delta winding, the c.t. in the star-point of the grounding trans-former (CT3 and CT4), and the high-impedance burdens R of the relays HIR1, respectively HIR2, are connected in parallel. Under normal load conditions, the voltage across the burden R is zero, because the geometrical sum of the phase currents measured by the Holmgren arrangement of the phase c.t's is zero and no current flows through the star-point to ground. Neglecting any transformer er-rors, the voltage across the resistor R is also zero during an external fault, since the differential (spill) current $\Delta I = I_\Sigma - I_E = 0$. The through-fault stability of the

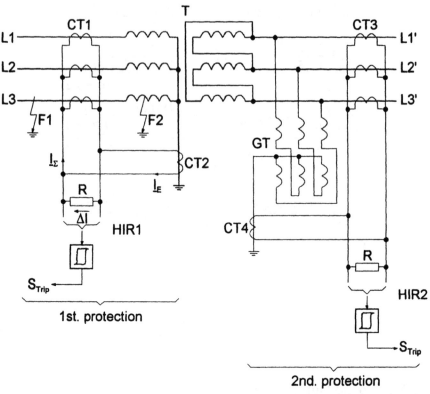

Fig. 7.40:   Restricted ground fault protection of a two-winding transformer using high-
             impedance relays
             GT = grounding transformer,   HIR = high-impedance relay,
             R   = input impedance of high-impedance relay

protection is achieved by setting a relay pick-up voltage which is higher than the
highest voltage generated across the circulating current circuit in the event of one
c.t. saturating [7.41].

   In the case of an internal fault on the other hand (fault F2), the entire secondary current flows through the resistor R and generates a high voltage across it
which is higher than the setting on the relay HIR1. A ground fault on the delta
windings is detected in a similar manner by the relay HIR2.

   Restricted ground fault protection is also used for detecting ground faults on
an auto-transformer. The corresponding circulating current connection of the
c.t's and the relay HIR is given in Fig. 7.41. In this case as well, the voltage
across the stabilizing resistor $U_R \approx 0$ for a through-fault (fault location F1), be-

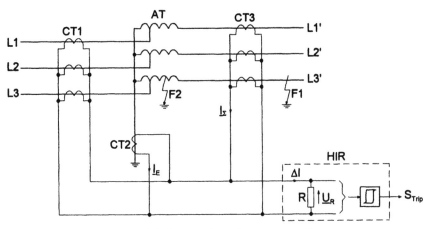

Fig. 7.41:   Restricted ground fault protection of an auto-transformer

cause $\Delta I \approx 0$. The situation is reversed during an internal fault (fault location F2) with all the current flowing through the stabilizing resistor such that $U_R > U_P$, U_P being the pick-up setting on the high-impedance relay.

Apart from the restricted ground fault protection schemes described above which only detect ground faults in the zone between the c.t's, i.e. on the transformer windings (hence the name "restricted"), ground fault schemes are also in use which detect internal and external faults. This applies to the simple scheme shown in Fig. 7.42 which only measures the ground fault current. Obviously this scheme can only detect ground faults on the star-connected winding and the part of the power system connected electrically to it. The relay setting is typically 20 % to 40 % of the rated transformer current which is adequate to protect the entire star winding. The operating time is chosen to grade with the longest time set on the ground fault protection relays in the next station.

Yet another very simple ground fault protection scheme for power transformers is the transformer tank protection. This requires that the mounting of the transformer tank insulate it from ground and that the tank then be grounded via a

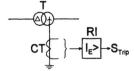

Fig. 7.42:   Protection against internal and external faults on the star winding of a two-winding transformer

c.t., the secondary of which is connected to an overcurrent relay. Since under normal load conditions no current flows through the c.t., the relay may have a very sensitive setting.

7.2.3 Buchholz protection

A Buchholz relay is a standard protection fitted to all oil-immersed transformers which detects all insulation breakdowns inside the transformer tank causing either the formation of gas or surges of oil flow from the tank to the expansion vessel. This applies to all phase and ground faults on the windings and to inter-turn faults. The relay also detects losses of oil caused by leaks as well as defects such as broken conductors and corroded or otherwise bad connections.

The operating principle of the Buchholz relay is based on the fact that firstly the pressure of gas in the upper part of the transformer tank increases due to the chemical decomposition of the oil and/or the combustion of solid insulating materials and secondly that massive gas development gives rise to a surge of oil towards the expansion vessel.

A Buchholz relay is installed in the connecting pipe between the transformer tank and the expansion vessel. It generally comprises two floats one above the other. The upper one signals the slow collection of gas (1st. stage) and the lower one a drop in oil level. A flow paddle is usually combined with the lower float which in the event of a surge of oil operates a mercury contact to trip the transformer.

According to the design of transformer, Buchholz relays are also used for the tap changer.

7.2.4 Overload protection

The purpose of the overload protection is to prevent overheating of the transformer due to inadmissibly high loads. A simple time-overcurrent relay can perform this function in the case of small transformers, with the disadvantage that it does not take the load history prior to the overload into account. Using an overcurrent relay is thus only an approximation and not a true overload protection. A principle is therefore employed for larger transformers based on a thermal image of the transformer. Such relays usually detect several defined temperatures and give alarm in stages to increase the forced cooling and initiate a reduction of load. The transformer is only tripped, if the overload persists for a long period or the temperature rise approaches its maximum permissible limit [7.42].

7.2.5   Integrated protection systems for transformers

As explained in Section 7.2.1, the choice of protection depends among other things on the rated power of the transformer. In the meantime it has been shown (Section 7.2.2) that it is also influence by the number of windings, method of grounding etc. There are therefore no standard uniform solutions and this Section can only point to a few examples which are used frequently in practice.

Figure 7.43a shows an integrated protection system for a small two-winding transformer with a rating of a few tens of MVA. The main protection for detecting phase faults is the biased differential protection (3) supported by the Buchholz relay (4). The time-overcurrent protection (1) serves as back-up and overloads are detected by the overload protection (2). A temperature measurement (5) guards against the oil exceeding its maximum permissible temperature. The corresponding tripping logic is given in Fig. 7.43b.

An example of a much more extensive system is shown in Fig. 7.44 for a large three-winding transformer (> 100 MVA). Internal faults are once again detected by a differential protection (6), the ground fault protection (1) (only star-connected winding) and the Buchholz relays (2) and (3). Two distance relays DR1 and DR2 and the time-overcurrent protection (8) are included for back-up protection. Overloading of any of the windings is signalled by the overload protection (7).

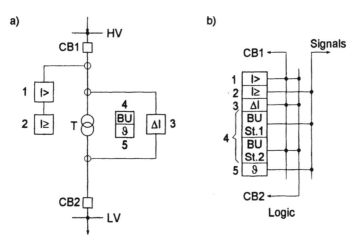

Fig. 7.43:   Integrated protection system for a small two-winding power transformer

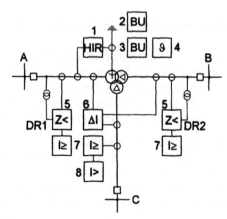

Fig. 7.44:   Integrated protection system for a large three-winding power transformer

Figure 7.45 similarly shows a protection system for a large auto-transformer for detecting all the faults and abnormal operating conditions which can possibly occur. The reaches of the two distance relays (4) generally correspond to those in Fig. 7.39. The differential relay (3) for phase faults can be either of the biased or especially in the case of an auto-transformer of the high-impedance type.

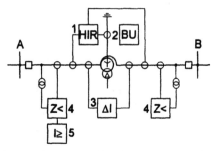

Fig. 7.45:   Integrated protection system for a large auto-transformer

7.3   Busbar protection and breaker back-up protection

7.3.1   Introduction

The term busbar protection refers to special protection systems designed exclusively to detect phase and ground faults in the bus zones of HV and EHV substations and switch yards. The purpose of breaker back-up protection is to clear a fault on a feeder by tripping all the other circuit-breakers on the same section of busbar should the feeder circuit-breaker fail to operate in response to a trip command. Since the busbar protection already includes a logic to enable it to isolate a specific section of busbar, busbar protection and breaker back-up protection are generally integrated in the same system.

Although the probability of a busbar fault is much lower than for other items of plant in a power supply system, when one occurs it can have serious consequences for the power system as a whole. This is because on the one hand the failure of the protection to clear an internal fault can cause extensive damage and disrupt power system stability, but it is equally disastrous on the other hand for it to mal-operate, because an important node is then removed which can also upset the operation of the entire power system. Busbar protection must therefore fulfil especially high requirements with respect to selectivity, operating speed and adaptability to changing busbar configurations.

A busbar protection system operates according to one of the following principles [7.43]
- current comparison with current restraint
- high-impedance
- phase comparison of currents
- logical evaluation of feeder protection devices

The following Sections describe in detail the measuring principles used for busbar protection and then the operation of breaker back-up protection.

7.3.2   Current comparison busbar protection

7.3.2.1 Current comparison with current restraint

The measuring principle corresponds basically to that of the biased differential protection (see Section 3.2, 6.2 and 7.2) with some adaptation to the particular problem and operating conditions.

The current restraint principle is shown in Fig. 7.46 for a single busbar. The number of feeders L has been reduced for the sake of simplicity to three. The operating current I_{op} is the geometric sum of all the secondary feeder currents and

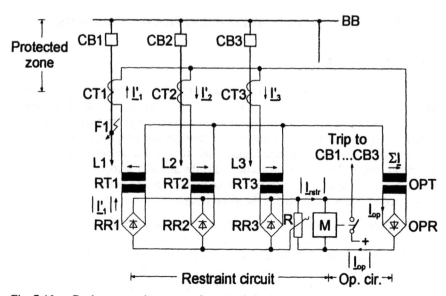

Fig. 7.46: Busbar protection system for a single busbar with current comparison and current restraint
M = measuring unit, RT = restraint interposing c.t., OPT = operating interposing c.t., RR = restraint rectifier bridge, OPR = operating rectifier bridge

the restraint current I_{rstr} is the arithmetic sum. The latter ensures that the protection remains stable during an external fault. Thus

$$I_{op} = \left|\sum_{n=1} I'_n\right| = |I'_1 + I'_2 + I'_3 + ... + I'_k| \tag{7.23}$$

$$I_{rstr} = k_{rstr}\sum |I'_n| = k_{rstr}\left(|I'_1| + |I'_2| + |I'_3| + ... + |I'_k|\right) \tag{7.24}$$

where

I'_n = c.t. secondary currents of the feeders $n ... k$
k_{rstr} = stabilization factor given by the ratio I_{op}/I_{rstr}

By appropriately setting the resistor R in the restraint circuit, the value of the stabilization factor and with it the slope of the operating characteristic in Fig. 7.47 can be chosen such that the protection remains stable during external faults even in the presence of extreme c.t. saturation. Typical settings for k_{rstr} lie between 0.5 and 0.9. Therefore during an external fault $I_{rstr} > I_{op}$.

Obviously the currents can only be compared, if the main c.t's CT of all the feeders L have the same ratio. Where this is not the case, the currents have to be balanced by adjusting the ratios of the interposing c.t.'s.

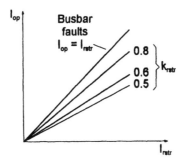

Fig. 7.47:   Operating characteristic of a busbar protection system with current restraint

For reasons of economy and simplicity, the phase currents of the individual feeders are frequently reduced to a single-phase measurement with the aid of mixing c.t's (see Section 5.3.1.2). The mixing c.t's can be connected in different ways to the main c.t's to achieve the desired performance for different sets of power system conditions. Of the possible alternatives, two predominate, the difference between them being normal or enhanced ground fault sensitivity. These two alternatives are shown in Fig. 7.48 [7.44].

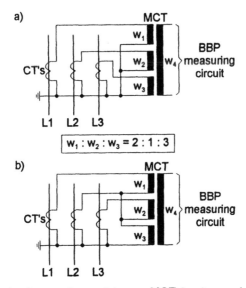

Fig. 7.48:   Methods of connecting a mixing c.t. MCT for a) normal and b) enhanced
              ground fault sensitivity [7.44]
              BBP   = busbar protection

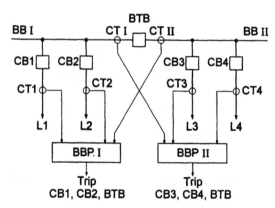

Fig. 7.49:   Protection of two sections of busbar connected by a bus-tie breaker

Where there are two sections of busbar connected by a bus-tie breaker, each section has its own protection system (Fig. 7.49). In order to maintain discrimination when the bus-tie breaker BTB is closed, the fault current conducted by the bus-tie breaker must be included in the measurement.

The protection for multiple busbars (double busbars with and without bypass busbar, 1½ breaker systems, triple and quadruple busbars) must be designed to automatically adapt its configuration to the configuration of the primary system. This is necessary so that the operating and restraint signals are derived separately for each separable section of busbar and also to ensure that the tripping signals only go to the circuit-breakers bounding the faulted section of busbar, i.e. only the smallest part of the busbar system surrounding the fault is isolated from the rest and all healthy sections remain unaffected. This mode of operation is achieved with the aid of a logic controlled by auxiliary contacts on all the busbar, bypass busbar and longitudinal isolators. The logic is thus a faithful replica inside the protection of the primary system. In certain circumstances, the current flowing across the transverse coupling must also be correctly included in the measurement.

The arrangement of the busbar protection and the allocation of the feeder currents to the busbar sections BBI and BBII in a double busbar configuration as determined by the locations of the isolators can be seen from Fig. 7.50. In this diagram the isolator circuits of the busbar replica are shown in a simplified form. In practice, the secondary feeder currents of the main c.t's CT are switched without interrupting the secondary circuits. Also the busbar replica is not built up using simple auxiliary relays, but electronic memories which record the "ON", respectively "OFF" positions of the isolators and retain the information even in the event of an auxiliary supply failure. The basic diagram of Fig. 7.50 is further simplified by omitting the tripping circuits and only indicating that the tripping

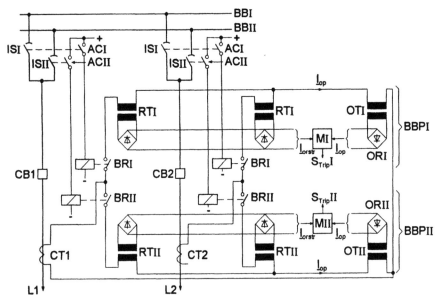

Fig. 7.50:   A differential busbar protection scheme with current restraint for a double
busbar configuration
ACI, ACII   = isolator auxiliary contacts on busbars I and II,   BRI, BRII
= busbar replica auxiliary relays,   MI, MII   = measuring systems for busbars I and II

signals $S_{Trip}I$ and $S_{Trip}II$ generated by the independent protection measurements
BBPI and BBPII are transmitted to the circuit-breakers switched to the respective busbars I and II.

Apart from the actual current summation circuit of the protection, there is an
additional measurement for each busbar section which continuously monitors the
integrity of the c.t. circuits. Should a spurious differential current occur due to
incorrect connection of the c.t. secondaries (short-circuit, reversed or open-
circuit), any tripping signal is blocked. The c.t. monitoring function is set to
about 10 % of the rated current defined for the busbar protection [7.43].

7.3.2.2 High-impedance busbar protection

The principle of this scheme was explained in part in Section 7.2.2.4. It is designed according to the differing conditions prevailing during internal and external faults while taking account of the saturation behaviour of the main impedance
of conventional c.t's. The measuring circuit represents a high impedance connected across the circulating current arrangement of all the c.t's connected in

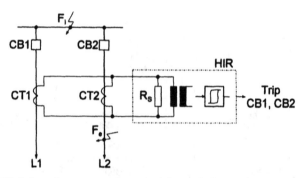

Fig. 7.51: High-impedance protection principle applied to busbar protection
HIR = high-impedance relay, R_S = stabilizing resistor

parallel. Figure 7.51 shows a phase and ground fault scheme for a single section of busbar simplified by reducing the number of feeders to just two. To facilitate proper comparison of the feeder currents in the circulating current circuit, all the c.t's must have the same ratio and also otherwise be generally of the same type and performance. According to the equivalent circuit of Fig. 7.52a and assuming the resistors R'_T and R''_T representing the c.t. winding and lead resistances to be equal, the voltage U_S across the high impedance R_S and the relay is zero under normal load conditions and during through-faults (fault F_e) without saturation of the c.t. CT2 on the feeder leading to the external fault. This is because the vectorial sum of the incoming and outgoing currents I_{1p} and I_{2p} is zero, i.e. the current "circulates" around the parallel circuit of the c.t's without any spill being left over to flow through the measuring circuit. X'_M and X''_M are the c.t. inductances.

For an internal fault on the busbar, the currents add instead of balancing each other and the combined current flows through the measuring circuit. The voltage across the circulating current circuit and the relay exceeds the setting of the protection and causes it to trip.

The main feature of a high-impedance scheme is its "through-fault stability", i.e. its non-response to c.t. saturation during an external fault, the probability of which is extremely high because the c.t. of the faulted feeder has to conduct the combined fault current contributions of all the other feeders connected to the busbar. When this happens, the reactance X_M of the outgoing c.t. is reduced to zero (Fig. 7.52b) and it ceases to supply a secondary current. The burden for the incoming c.t's now comprises the high impedance of the measuring circuit in parallel with the relatively lower impedance of the leads from the relay location to the saturated c.t. and the resistance of the saturated c.t. winding. The value of the stabilizing resistor R_S is chosen such that voltage drop across the saturated c.t. and its leads cannot reach the pick-up voltage of the measuring circuit. The maximum voltage U_R at the relay location is given by

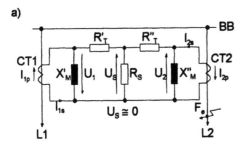

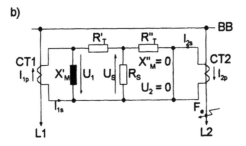

Fig. 7.52:   Effect of the stabilizing resistor during an external fault
              a) normal load conditions or external fault without saturation
              b) through-fault with complete saturation of the outgoing c.t. in the line LZ2

$$U_R = I_{smax}R''_T \tag{7.25}$$

where
I_{smax} = maximum secondary through-fault current

To ensure that there is no risk of the relay false tripping, the pick-up setting of the relay/stabilizing resistor combination is chosen to satisfy [7.41]

$$U_{set} \geq 2.5 I_{smax}(R_W + 2R_L) \tag{7.26}$$

where
R_W = resistance of the c.t. secondary winding
R_L = resistance of the longest c.t. lead from the paralleling point/relay branch off

It is thus important to keep the lead resistance as low as possible and to select c.t's with a low winding resistance and negligible leakage flux. It is also necessary to know the knee-point voltage of the c.t.'s in order to be able to design the scheme and determine its sensitivity. To limit the maximum voltage across the relay and the entire parallel circuit of the c.t.'s during an internal fault when all the

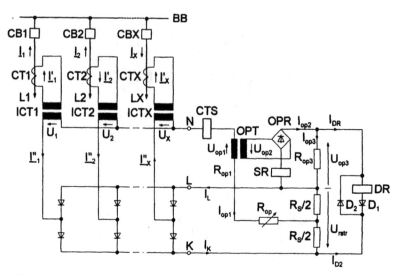

Fig. 7.53:   Improved high-impedance protection according to [7.45]
             SR = starting relay,   DR = differential relay,   CTS = c.t. circuit su-
             pervision relay

feeder currents flow in the same direction, a practical scheme also includes voltage limiting components across the parallel circuit of the c.t's and the relay.

The traditional high-impedance scheme described up to the present is only
really suitable for single busbars in MV installations. The improved version of
this scheme described below can be applied to either single or multiple busbars
[7.45]. Basically it is the combination of a normal high-impedance scheme and a
current stabilized differential scheme. Its single line diagram for a single busbar
with the feeders L1, L2 and LX is shown in Fig. 7.53. In this scheme, the restraint voltage U_{rstr} is proportional to the restraint current and the operating voltage U_{op3} to the operating current I_{op} (differential current).

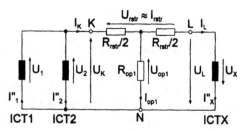

Fig. 7.54:   Simplified equivalent circuit of the protection scheme in Fig. 7.53 during an
             internal fault [7.45]

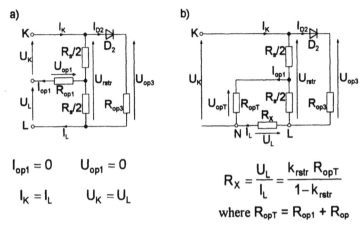

$$I_{op1} = 0 \qquad U_{op1} = 0$$

$$I_K = I_L \qquad U_K = U_L$$

$$R_X = \frac{U_L}{I_L} = \frac{k_{rstr}\, R_{opT}}{1 - k_{rstr}}$$

where $R_{opT} = R_{op1} + R_{op}$

Fig. 7.55:   Voltage distribution in a stabilized high-impedance scheme for a) an external fault a) without c.t. saturation and b) with complete c.t. saturation
[7.45]

Neglecting the diodes in the various feeders, the measuring circuit can be replaced by the equivalent circuit of Fig. 7.54 for the normal load condition or a
through-fault on feeder LX which is not accompanied by c.t. saturation. The corresponding measured variables are shown in Fig. 7.55a. The total incoming current I_K flows into terminal K of the relay and the total outgoing current I_L from
terminal L. In the idealized case, i.e. neglecting c.t. errors, $I_K = I_L$ and the operating (differential) current $I_{op1} = 0$. The restraint voltage U_{rstr} is generated between
the terminals K and L and drives a current I_{D2} through the diode D_2 and the resistor R_{op3} to the output terminal L. Under these conditions, operation of the differential relay DR (see Fig. 7.53) is blocked and the protection does not pick-up.

Assuming the c.t. CTX in feeder LX completely saturates during a through-
fault, the conditions for the protection change to those illustrated in the simplified
equivalent circuit of Fig. 7.55b. The total resistance R_x looking from terminal L
towards the feeder LX now lies between terminals L and N. The current I_L produces the voltage drop U_L across this resistor. The protection will not operate,
i.e. the relay DR (Fig. 7.53) will not pick up, providing the following condition
corresponding to the characteristic for restrain is fulfilled

$$R_x \le \frac{k_{rstr} R_{opT}}{\left(1 - k_{rstr}\right)} \qquad (7.27)$$

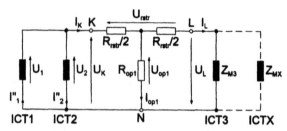

Fig. 7.56:   Simplified equivalent circuit of the scheme shown in Fig. 7.53 for a fault on
the busbars [7.45]

where
$$R_{opT} = R_{op1} + R_{op}$$

From this equation and equation

$$I_{op1} = k_{rstr} I_k \tag{7.28}$$

it follows that the magnitude of the fault current has no influence on the stability of the protection.

The conditions for an internal fault are given in Fig. 7.56, assuming that the busbar fault can only be supplied with fault current via feeders L1 and L2 and the other feeders (L3 to LX) are either not energized or are not on load. Accordingly, the impedances of the c.t's of all the unloaded feeders are connected to terminal L

$$Z_L = \frac{U_L}{I_L} = Z_{LM} \tag{7.29}$$

During an internal fault, the impedance Z_{LM} is normally much greater than the resistance R_{op1} and it is for this reason that the current I_{op1} is also greater than I_L, which ensures reliable operation of the differential relay DR in a time of 1 to 3 ms.

The measuring circuit includes two current relays in addition to the differential relay DR to increase the reliability of the protection. One is a starting relay SR and the c.t. circuit supervision relay CTS (see Fig. 7.53). Both relays perform a similar function, i.e. to prevent mal-operation of the protection should a c.t. become open-circuit.

7.3.3 Phase comparison busbar protection

The measuring principle of this scheme is shown in a simplified form for a fault on the busbar in Fig. 7.57 and for a through-fault in Fig. 7.58. The diagrams assume that none of the c.t's saturate and that the currents I_1 to I_3 are in phase for a busbar fault, i.e. that the feeder impedances are largely the same. A further simplification is that the comparison arrangement is only shown for the positive half-cycles of the phase currents i_1 to i_3. The same measurement is also performed for the negative half-cycles in a practical scheme.

It can be seen from Fig. 7.57 that the generation of the enabling signal ES depends on the coincidence time t_c, i.e. the time during which all the positive half-cycles, respectively the squarewave signals I_{1R} to I_{3R} derived from them, overlap. Since the coincidence time (maximum $t_c = 8.33$ ms for $f_N = 60$ Hz) is sufficiently long for an internal fault (F1), the enabling signal permitting tripping is generated. For a fault which does not lie within the protected zone (F2 in Fig. 7.58), the phase-angle of the current in the feeder with the fault is at 180° in relation to the currents of the other feeders, i.e. overlapping of all the current half-cycles does not occur and the enabling condition is not fulfilled.

In view of the different transmission line impedances, c.t. errors, the DC component of the fault current, but mainly because of the virtually unavoidable saturation phenomenon especially during through-faults, precautions have to be added to the basic principle of the scheme to ensure its correct operation in prac-

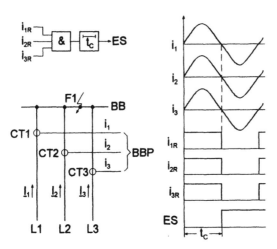

Fig. 7.57: Principle of a phase comparison busbar protection scheme for an internal busbar fault [7.46]

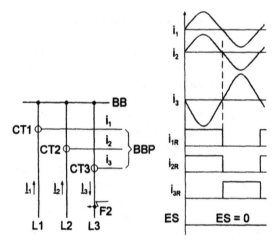

Fig. 7.58: Principle of a phase comparison busbar protection scheme for an external fault [7.46]

tice. The configuration of a practical scheme is shown in Fig. 7.59 [7.46]. Three conditions have to be fulfilled before tripping can take place

1. The duration of coincidence t_{cP} and t_{cN} for the positive and negative half-cycles of all the feeder currents must be longer than t_1 where $t_1 > 5$ ms (enabling signal 1).

2. The duration of non-coincidence t_{ac} must be less than t_2 where $t_2 < 3.33$ ms (blocking signal BS).

3. At least 1 ... 2 ms of coincidence must be measured for the next half-cycle of opposite polarity (enabling signal 2).

In cases of extreme c.t. saturation, it is possible for tripping by a phase comparison scheme not to be enabled for a busbar fault in spite of the methods and criteria employed. This possibility can, however, be avoided by using the first derivation of the currents di_1/dt ... di_x/dt instead of the currents i_1 ... i_x [7.46].

7.3.4 Directional current comparison busbar protection

A current comparison function is frequently added to a phase comparison function to increase the reliability of a busbar protection scheme which is then referred to as directional current comparison. The principle of this arrangement is shown in Fig. 7.60 for a simplified busbar of just two feeders. Phase comparison is performed between the secondaries of the interposing c.t's ICT-PC and current comparison between the secondaries of the interposing c.t's ICT-CC and the

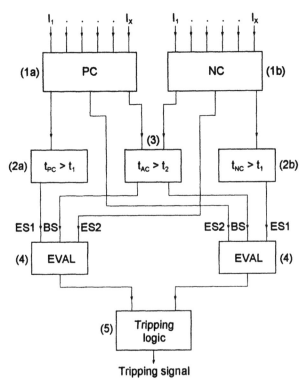

Fig. 7.59:   Complete phase comparison scheme for busbar protection [7.46]
        1a   = phase comparison of positive half-cycles
        1b   = phase comparison of negative half-cycles
        2a, 2b = coincidence time check
        3   = non-coincidence time check
        4   = evaluation, ES = enabling signal, BS = blocking signal

matching c.t. MCT in the differential circuit. Once again there are independent measuring systems for positive (P) and negative (N) half-cycles. High security against false tripping is achieved by logically relating the two criteria in a NAND gate.

Yet another alternative is based on the simultaneous detection of the following variables

- amplitude and phase-angle of the differential current
- amplitude and phase-angle of the stabilization factor k_{rstr}
- a saturation blocking signal as directional criterion for through-faults

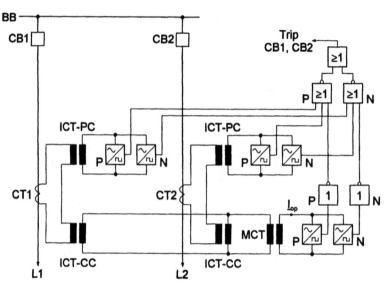

Fig. 7.60: Principle of directional current comparison busbar protection [7.47]
ICT-PC = interposing c.t's for phase comparison
ICT-CC = interposing c.t's for current comparison
MCT = matching c.t. in the differential circuit

The derivation of the saturation blocking signal is explained with reference to the diagram of Fig. 7.61 and the waveforms for the positive half-cycle for one feeder given in Fig. 7.62 [7.48, 7.49]. The main c.t. secondary current is appropriately adjusted by the interposing c.t. ICT. The voltage U_R has the same waveform and phase-angle and the voltage U_c across the condenser C increases as the current increases. At the zero-crossing of the current i_s, $U_c = U_D$ due to the condenser discharging, U_D being the breakdown voltage of the diode D. The saturation signal generator SSG forms a differential voltage U_Δ from the voltages U_c and U_R (curve 4). Saturation of the c.t. CT causes the pick-up level U_A (proportional to the primary current $I_p = 1$ kA) to be exceeded for a time in excess of 2.2 ms (see Fig. 7.62b). This also influences the blocking signal BS-P, the duration of which increases with increasing degree of saturation. This also applies for the negative half-cycles which are not shown in Fig. 7.62. The saturation blocking and other protection signals are interlocked such that tripping can only take place when the following conditions are fulfilled
- The differential current must exceed the setting I_{Fmin}.
- The instantaneous value of the stabilizing factor must be greater than its setting.
- Both the above criteria must be fulfilled simultaneously for at least 4.16 ms.
- The saturation blocking signal BS must not block during the 4.16 ms.

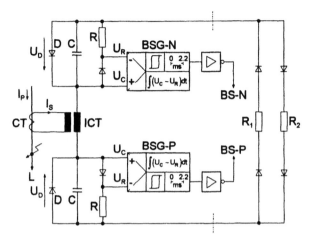

Fig. 7.61:   Derivation of the saturation blocking signal BS [7.48]
BSG-P, BSG-N = blocking signal generators for positive and negative
half-cycles

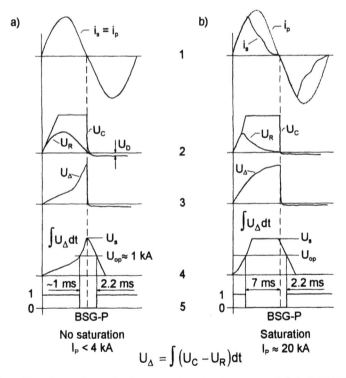

$$U_\Delta = \int (U_C - U_R) dt$$

Fig. 7.62:   Signal waveforms for the saturation signal generator BSG-P [7.48]

7.3.5 Busbar protection by linking overcurrent protection devices

A simple protection can be built up for busbars in MV installations having only a single infeed using the feeder overcurrent relays. To keep tripping times for busbar faults to a minimum, this requires interlocking the relays on the incoming feeder by the relays on the outgoing feeders. A corresponding logic is shown in Fig. 7.63. The overcurrent relay on the incoming feeder has two operating times t_G and t_{BB}. t_G is set to grade with the other relays in the system, but t_{BB} does not have to grade with the other relays and is set much shorter, usually only a few tens of milliseconds, i.e. the circuit-breaker CBI is tripped after the time t_{BB} for a fault on the busbars. A fault outside the bus zone on one of the outgoing feeders L1 or L2 causes the corresponding relay to pick up and transmit an undelayed blocking signal BS via the OR gate to block fast tripping by the incoming feeder relay and the incoming feeder breaker CBI is not tripped. Normally the external fault will be cleared after either time t_1 by tripping CB1 or time t_2 by tripping CB2 depending on the location of the fault. Should the respective protection or circuit-breaker fail to clear a fault on an outgoing feeder, the protection on the infeed steps in in a back-up function and trips the circuit-breaker CBI after the grading time t_G.

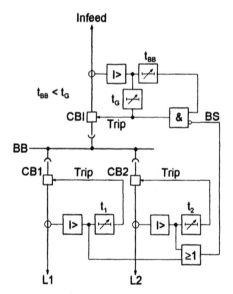

Fig. 7.63:   Busbar protection by interlocking feeder overcurrent relays

7.3.6 Breaker back-up protection

The circuit-breaker is the last link in the chain of protection, but in contrast to the other links in the chain (c.t's, v.t's, relays, battery etc.) it is not duplicated because of the high cost. Thus should a circuit-breaker fail to operate, a back-up system is needed which is able to clear the fault by tripping the immediately neighbouring circuit-breakers on the same busbar and possible also the circuit-breaker at the other end of the effected line. The protection scheme which performs this function is called breaker back-up protection [7.50, 7.51, 7.52].

A breaker back-up protection scheme (BBU) monitors the time which expires between the generation of a tripping signal by the feeder protection and the discontinuation of the fault current. Figure 7.64 shows in a simplified form the co-ordination of the BBU protection with the feeder protection PD (e.g. distance protection) and Fig. 7.65 the operating sequence for a) correct operation of the circuit-breaker CB3 for a fault on feeder L3 (fault at F) and b) failure of the cir-cuit-breaker to execute the tripping command. Important for the proper function of the BBU is that the current unit BBUP must have a very short reset time (< 20 ms) even in the presence of c.t. saturation phenomena.

From Fig. 7.64 it can be seen that when the protection device PD trips for a fault on feeder L3, it transmits a signal to the AND gate of the BBUP in addition to the tripping signal to the circuit-breaker. Providing the circuit-breaker responds to the tripping command, the current I_{BBU} becomes zero and the timer t_{BBU} is not started. In the event that the circuit-breaker should fail to open, the current I_{BBU} continues to flow, the current unit I> remains picked up and the timer is started. After about 100 ms, the BBUP issues tripping commands to the

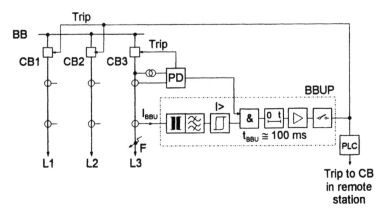

Fig. 7.64: Principle of breaker back-up protection (BBU)
PD = protection device on feeder L3

a)

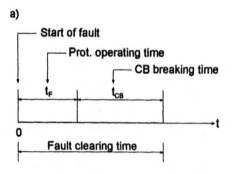

b)

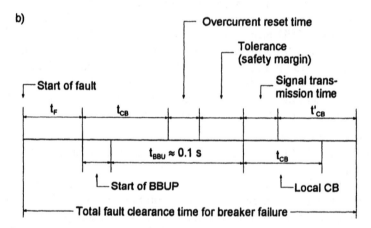

Fig. 7.65: Operating sequence for a) an intact circuit-breaker and b) a defective circuit-breaker

t'_{CB} = operating time of the circuit-breaker in the remote station

other circuit-breakers connected to the same busbars (CB1 and CB2) and also to the circuit-breaker at the opposite end of feeder L3.

It is modern protection practice to integrate the breaker back-up protection in the busbar protection, because it uses functions which are already available in the busbar protection. This is especially so in multiple busbar installations, in which both busbar and breaker back-up protections can use a common busbar replica logic controlled by auxiliary contacts on the isolators. This ensures that the breaker back-up protection also only isolates the smallest part of the system when clearing the fault following the failure of a circuit-breaker. In such combined systems, the breaker back-up protection achieves its objective by simply inhibiting the measurement of the outgoing feeder current (feeder with the failed breaker) by the busbar protection, so that a busbar fault is simulated and the busbar protection automatically trips the correct set of circuit-breakers [7.46, 7.48].

8

Analogue Protection for Machines

The discussion in this chapter is confined to the selective protection of the following kinds of electrical plant:

- synchronous generators
- generator/transformer units
- three-phase HV motors

Special attention is paid to those protection devices which have not been dealt with at all or only to a limited extent up to the present.

8.1 Generator protection

8.1.1 Kinds of faults and selective protection schemes

The synchronous generator is indisputably the most important item of plant in an electrical power system and for this reason its protection will be gone into in some detail with special reference to large generators [8.1...8.5].

A much wider variety of faults can befall a synchronous generator than any other item of plant on the power system. These can afflict its stator or rotor windings and be the result of either an internal breakdown of the insulation or an external condition on the power system or in the excitation circuit. Table 8.1 lists

Table 8.1   Faults which can occur on synchronous generators

Internal faults	External faults
Stator Ground faults Phase faults between windings Inter-turn faults **Rotor** Ground faults Duplicate ground faults	Phase faults Asymmetric load (NPS) Loss of excitation Pole slipping Stator overload Rotor overload Overvoltage Underfrequency Motoring *)

*) a hazard for steam turbines

those kinds of faults which can be detected through the supervision of electrical system variables.

The amount of protection installed depends on a number of factors, the most important of these being the rated power of the generator and the ratio of its power to the total power of the system. Others are the configuration of the primary plant, i.e. whether the machine is directly connected to the busbars or via a step-up transformer, method of star-point grounding and the type of excitation. Of consequence is also whether the generator is in a coal-fired, nuclear, hydro-electric or pump storage power plant [8.6...8.10]. For these reasons and also because the opinions, recommendations and guidelines vary not only from one country to the next, but also from one utility to the next, it is virtually impossible to achieve complete standardization regarding the amount and types of protection for generators and generator/transformer units [8.11, 8.12]. The same applies to the precautions against underfunction (fails to operate when it should) and overfunction (operates when it should not) of the protection and to some degree to the associated switchgear or the initiation of other measures (e.g. reducing the excitation).

The introduction of solid-state protection devices and modular system structures (see Chapter 5) has made it possible to adapt the generator or generator/transformer unit protection to satisfy the widely varying demands. A parallel development has also been the division of the protection into two completely independent groups, each with its own auxiliary supply source, to achieve redundant protection systems. The addition of permanently installed automatic test equipment was a further innovation to enhance protection availability [8.13, 8.14].

The following sections of this chapter explain the operating principles of the protection devices and schemes currently in use for detecting the kinds of faults listed in Table 8.1.

8.1.2   Stator ground fault protection

A stator ground fault is one of the most frequent internal generator faults which is caused by physical damage to the stator winding or aging of its insulation. Since the star-point of a generator is generally not directly grounded, stator ground fault currents are low, especially in the case of generators of a generator/transformer unit, but even low ground fault currents can cause considerable damage to the iron core or evolve into a phase fault. Stator ground faults must therefore be detected and the generator shut down in as little time as possible.

The method chosen to detect stator ground faults depends essentially on how the generator is connected to the power system. A basic distinction is made between
- generators without a step-up transformer which are directly connected to the busbars
- generators with a step-up transformer (generator/transformer units or sets)

Generators which are directly connected to the busbars are typically of low power and the stator ground fault protection is based on supervising either the

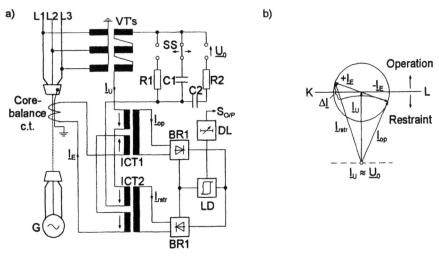

Fig. 8.1:   Basic circuit a) and corresponding vector diagram b) for a directional stator ground fault scheme for generators connected directly to the busbars [8.15]

amplitude [8.5] or the direction [8.15, 8.16] of the neutral current. The neutral current is measured using either a core-balance c.t. or three single-phase c.t's in a Holmgren arrangement. Monitoring the amplitude is applicable in ungrounded systems where the current I_{ES} flowing from the system to a ground fault in the generator is much larger than the current I_{EG} flowing from the generator to a ground fault on the external system, i.e. the condition $I_{ES} \gg I_{EG}$ is fulfilled. The zone of protection extending from the generator terminals towards the star-point for a scheme of this kind is limited to approximately 70 % of the stator winding.

A directional ground fault scheme extends the zone of protection to approximately 90 %. The corresponding principle is shown in Fig. 8.1 [8.15]. The variable supervised in this case is ΔI (Fig. 8.1b) which is described by the following equation

$$\Delta I = \left| \underline{I}_{rstr} \right| - \left| \underline{I}_{op} \right| \tag{8.1}$$

where

$$\underline{I}_{rstr} = \underline{I}_U + \underline{I}_E \tag{8.2}$$

and

$$\underline{I}_{op} = \underline{I}_U - \underline{I}_E \tag{8.3}$$

In the above equations, $\underline{I}_U$ is proportional to the voltage off-set $\underline{U}_0$ (Fig. 8.1a) and $\underline{I}_E$ is the neutral current provided by the core-balance c.t. It can be seen from the vector diagram that the condition for tripping is determined by the sign of ΔI, tripping taking place when it is positive, i.e. for $\left| \underline{I}_{rstr} \right| > \left| \underline{I}_{op} \right|$. This condition is fulfilled by a stator ground fault in the zone of protection. The horizontal line K-L in Fig. 8.1b thus marks the border between operation and restraint.

If the resistor R_1 is connected across the broken-delta v.t. secondaries in place of the capacitive burden C_1, the above scheme can also be used for protecting a generator grounded by a high resistance. The real power component of the current is then evaluated compared to the reactive power component when the star-point is ungrounded. If the resistive and capacitive components of the current are approximately equal, the so-called 45° circuit is used which is achieved by selecting the mixed circuit of C_2 and R_2 in series as the burden across the broken-delta of the v.t's.

An alternative arrangement for detecting ground faults on the stators of generators which have ungrounded or high-resistance grounded star-points and are directly connected to the busbars is shown in Fig. 8.2 [8.16].

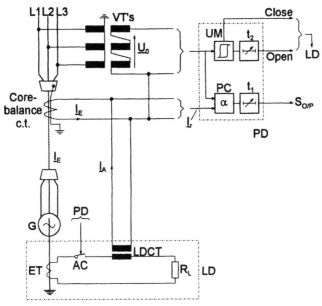

Fig. 8.2:   Basic circuit of a directional stator ground fault protection scheme for generators directly connected to the busbars

The difference compared to the scheme described above is that regardless of whether the star-point is grounded or not, a star-point loading device LD is activated upon the detection of a ground fault which artificially increases the real power component of the current to about 10 A to facilitate correct directional measurement. The loading device consists of a grounding transformer ET with a loading resistor R_L connected across its secondary via an auxiliary contact AC. The existence of a ground fault is detected by the voltage measuring device UM which responds to the appearance of the off-set voltage U_0. The auxiliary contact AC closes immediately for a time determined by the setting t_2. Providing the ratio of the c.t. LDCT in the loading device is correctly chosen, sufficient active power now flows to enable the phase comparator PC to clearly determine the fault direction. Figure 8.3 shows a simplified equivalent circuit and the corresponding vector diagrams for ground faults outside (Fig. 8.3a) and inside (Fig. 8.3b) the zone of protection. It follows that the resultant current I_r applied to the phase comparator for an external ground fault is the difference between the real power current I_A caused by the loading resistor R_L and the ground fault current I_E, i.e.

$$I_r = I_A - I_E \tag{8.4}$$

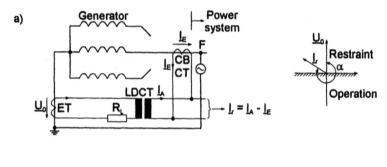

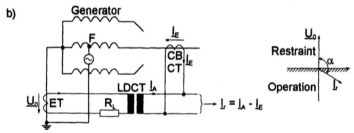

Fig. 8.3:   Simplified equivalent circuit and corresponding vector diagrams for a stator
               ground fault protection scheme which adds a real power component to the
               fault current [8.16]
               a) ground fault outside the zone of protection
               b) ground fault inside the zone of protection

In this case, the phase-angle α between the off-set voltage $\underline{U}_0$ and the resultant ground fault current $\underline{I}_r$ is greater than the limit angle for operation of the
phase comparators PC (Fig. 8.2) and no tripping signal is produced. For a
ground fault on the stator winding of the generator, the relationship becomes

$$\underline{I}_r = \underline{I}_A + \underline{I}_E \tag{8.5}$$

and the phase-angle α reduces to a value safely within the operating area.

The scheme described above achieves a zone of protection of 90 %, i.e. the
last 10 % close to the star-point are not protected. Since the maximum voltage to
ground for faults in this zone is only 10 % of the generator rated voltage, the
probability of a ground fault being caused electrically is low and therefore small
generators are not usually equipped with stator ground fault protection covering
100 % of the winding.

In the case of a generator/transformer unit, the detection of stator ground
faults is theoretically quite simple, because the primary of the step-up transformer
is always delta-connected and the transformer forms a natural barrier for ground

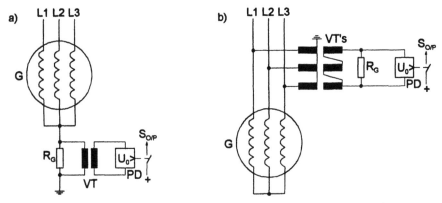

Fig. 8.4:   90 % stator ground fault protection for generator/transformer units
            a) measurement of the off-set voltage at the star-point
            b) measurement of the off-set voltage with the aid of a broken-delta arrangement of p.t's

faults on the high-voltage side. Thus ground faults on the stator windings and on all items of plant electrically connected to them can be detected by monitoring the 50/60 Hz off-set voltage between the generator star-point and ground. Figure 8.4 shows to alternative protection schemes based on this principle. The purpose of the resistors R_G is to limit the influence of the capacitance between primary and secondary of the step-up transformer on the transformation of the neutral voltage and therefore on the magnitude of the fault current for a ground fault on the HV power system [8.17]. The disadvantage of the scheme is that the zone of protection is only about 90 to 95 % of the stator winding. For this reason, large units are fitted with an additional 100 % stator ground fault scheme, which is considered necessary, because faults can occur close to the star-point as a consequence of mechanical damage such as creepage of the conductors and loosening of bolts [8.18]. Of the possible technical methods of achieving protection over 100 % of the stator windings, only the two most favourable alternatives are dealt with below [8.19]. These are based on the following principles
- supervision of the third harmonic in the voltages at the star-point and at the generator terminals
- injection of a low-frequency bias voltage at the generator star-point

   The first of the two measuring principles assumes that a considerable third harmonic component exists in the generator phase voltages due to unavoidable non-linearities in the magnetic circuits of the various generator designs. Under normal operating conditions, the distribution of the third harmonic along the stator windings corresponds to Fig. 8.5a [8.19]. It can be seen that the maxima occur at the star-point N and at the terminal T. The values increase with generator load. For a stator ground fault at the star-point (Fig. 8.5b), the amplitude of

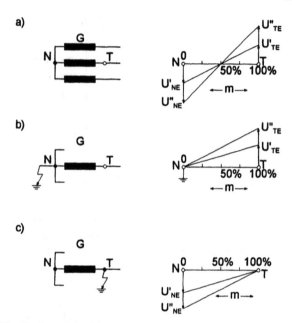

Fig. 8.5:   Distribution of the third harmonic component along the stator windings of a
            large generator [8.19]
            a) normal operation   b) stator ground fault at the star-point   c) stator
            ground fault at the terminals
            m = relative number of turns

the third harmonic in the voltage at the terminals is approximately doubled both
when the generator is off-load prior to the fault (U'_{KE}) and when it is fully loaded
(U''_{KE}). The same third harmonic values can be measured in the star-point voltages U'_{NE} and U''_{NE} for a stator ground fault at the generator terminals (Fig.
8.5c).

The protection scheme for detecting stator ground faults near the star-point
makes use of the above facts and compares the third harmonic contents of the
voltages at the star-point (U_{NE}) and at the generator terminals (U_{KE}). Figure 8.6
shows a simplified diagram of a 100 % stator ground fault protection which operates according to the above principle [8.20]. Protection relay RU2 is connected
to terminals e-E of a balanced voltage circuit via a filter tuned to the third harmonic. The impedances Z_1 and Z_2 of the balanced voltage circuit are chosen such
that in normal operation $\underline{U}_{eE} = 0$ (Fig. 8.7a), but assumes an appreciably higher
value for a stator ground fault close to the star-point (Fig. 8.7b).

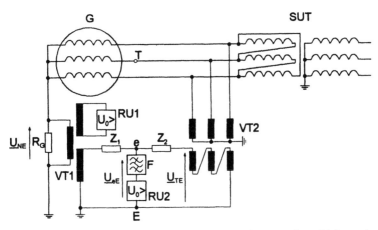

Fig. 8.6:   Basic diagram of a 100 % stator ground fault protection which monitors the
             third harmonic [8.20]

Both theoretically and practically, a stator ground fault protection scheme
based on the comparison of third harmonic contents is subject to certain short-
comings, e.g.

- the third harmonic content in the generator voltage is much reduced for a
  ground fault at the center of the winding (see Fig. 8.5a) and the protection
  cannot pick up.
- the value of U_{eE} is reduced by a high fault resistance [8.21].
- the scheme cannot detect ground faults when the machine is not running.
- the low third harmonic content of certain generator designs makes fault detection difficult.

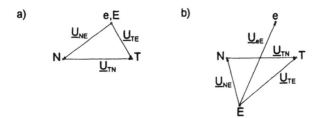

Fig. 8.7:   Vector diagrams of the third harmonic content of generator and protection
             device voltages [8.20]
             a) normal operation   b) ground fault close to the star-point

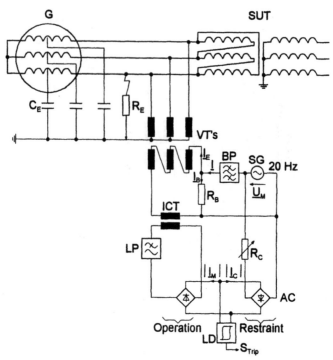

Fig. 8.8:   Basic diagram of a 100 % stator ground fault protection scheme with injection of a 20 Hz signal [8.23]

For these reasons, the use of the third harmonic to detect stator ground faults along 100 % of the stator windings is somewhat disputed, especially as in the meantime schemes have been developed which can do the same even better in all operating modes including when the set is at standstill and when starting up and shutting down. These are based on the second alternative of injecting a signal at the star-point at a frequency below the power system frequency. The basic diagram of a scheme of this kind is show in Fig. 8.8 [8.22, 8.23]. The injection voltage $\underline{U}_M$ from a signal generator (20 Hz for 50 Hz generators and 24 Hz for 60 Hz generators) displaces the generator star-point with respect to ground and causes a total 20/24 Hz current $\underline{I}$ to flow. This current divides between the fault resistance R_E and the grounding resistor R_G and only the portion of current flowing through R_E is applied to the measuring circuit AC via the interposing c.t. ICT and the low-pass filter LP. During normal operation ($R_E = \infty$), a small 20/24 Hz current I_M flows as determined by the capacitances C_E between the stator windings and ground. This quiescent current is compensated in the amplitude comparator AC by a current I_C which is also derived from the 20/24 Hz generator such that in normal operation $I_C > I_M$. A stator ground fault reduces the value of R_E and in consequence increases the measurement current I_M. Since the compensation current I_C does not change, $I_M > I_C$ and the limit detector LD picks up and produces and output signal. The inclusion of the low-pass filter in the operating part of the amplitude comparator means that the measured current I_M only represents the 20/24 Hz component and the 50/60 Hz component of the fault current is rejected. The purpose of the bandpass filter BP, on the other hand, is to protect the 20/24 Hz signal generator SG from being overloaded by the 50/60 Hz off-set voltage which occurs for a ground fault close to the generator terminals.

An alternative injection scheme is shown in a simplified form in Fig. 8.9 [8.18, 8.24]. In this case, a coded 12.5/15 Hz (for 50 Hz and 60 Hz machines respectively) current I'_{SH} generated by SG is injected into the primary circuit via an injection transformer IT. During normal operation this current only flows through the leakage capacitances C_{Eg} of the electrically connected system (Fig. 8.10a). In the event of a stator ground fault, the ground fault resistance R_E is in parallel with the total capacitance C_{Eg} (Fig. 8.10b) and the injection current I'_{SH} increases such that $I''_{SH} > I'_{SH}$. Operation of the level detector LD is thus determined by two conditions, the increase in the level of the injection current and the detection of the injected code in the signal coming back. Coding of the injection signal is achieved by injection periods and injection intervals as illustrated in the timing diagram of Fig. 8.11. Several measurements are carried out during transmission and during the intervals, measurements M_1, M_2 and M_3 being evaluated during transmission and P_1 to P_6 during the intervals. This technique avoids any risk of mal-operation due to spurious signals in the primary system current and involves the separate integration of positive and negative half-cycles to eliminate interference at frequencies which are multiples of 12.5/15 Hz.

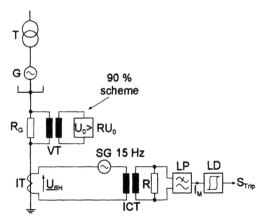

Fig. 8.9:   Simplified basic diagram of a 100 % stator ground fault protection scheme with injection of a coded signal at 15 Hz [8.24]

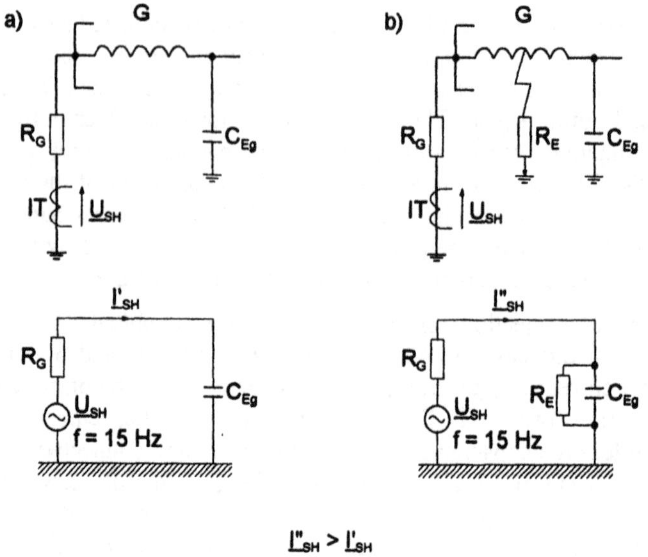

$\underline{I}''_{SH} > \underline{I}'_{SH}$

Fig. 8.10:   Simplified equivalent circuit of a stator ground fault protection scheme according to Fig. 8.9
a) normal operation   b) stator ground fault

A 100 % stator ground fault protection scheme is always applied in combination with an independent 90 % scheme as shown in Fig. 8.9.

8.1.3   Phase faults between windings

8.1.3.1 Kinds of protection

On all units with ratings of a few MVA and above, the main protection used today for detecting phase faults between the windings of synchronous generators is a biased differential protection scheme. Time-overcurrent relays are applied as back-up protection on small to medium sized generators and distance protection on large generators. The back-up protection also detects faults outside the generator zone.

8.1.3.2 Biased differential protection

The operating principle, the operating characteristic and a solid-state version of a biased differential protection scheme for electrical machines were described in detail in Section 6.2 so that here it is only necessary to add the special aspects applicable to generator differential protection.

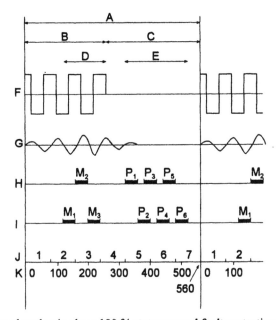

Fig. 8.11:   Signal evaluation by a 100 % stator ground fault protection scheme with
injection of a coded signal [8.24]
A = total measuring period,   B = transmission period,   C = interval,
D = measurement period,   E = check,   F = coded signal,   G = measured
signal I_M at the output of the low-pass filter LP,   H = positive half-cycle
integration times,   I = negative half-cycle integration times,   J = 15 Hz
periods,   K = time [ms]

The biased differential protection of a synchronous generator compares the
phase currents I_1 flowing at the star-point with the phase currents I_2 flowing in
the generator feeder (Fig. 8.12) as they appear at the secondaries of the c.t's CT1
and CT2. Figure 8.12 only shows the secondary circuit for one phase and a
complete scheme would have identical circuits for the other two phases. The
measuring unit M can perform the amplitude comparison of the operating current
I_{op} and the restraint current I_{rstr} in a number of different ways, of which one example is shown in Fig. 6.12 in Section 6.2. Schemes do exist which only have a
single measuring unit for all three phases [8.25], but as always with such schemes
the operating characteristic and therefore also the sensitivity vary according to
the phases involved in the fault.

Since in the case of generator protection the c.t's at the star-point and the
terminals have the same ratio, an alternative differential scheme for a generator is
high-impedance protection. Figure 8.13 shows the measuring conditions for a

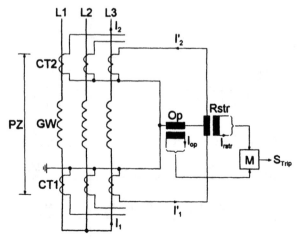

Fig. 8.12:   Principle of a circulating current circuit as applied in a biased differential
scheme for a generator
PZ = zone of protection

high-impedance scheme during normal operation (Fig. 8.13b) and during an external phase fault with complete saturation of CT2 (Fig. 8.13c). It follows from Fig. 8.13b that assuming the c.t's to be identical, i.e. with identical main inductances ($L_{m1} = L_{m2}$), the voltage across the high resistance R_D in normal operation is only dependent on the values of the resistors R_1 and R_2. $U_D = 0$ for $R_1 = R_2$. Resistors R_1 and R_2 represent the sums of the secondary resistances of CT1 and CT2 and the loop resistances of the leads between the relay and the c.t's on the respective sides of the circulating current circuit.

Should one of the c.t's, e.g. CT2, saturate completely during an external fault (F in Fig. 8.13a), L_{m2} becomes zero and therefore also its e.m.f. E_2 (Fig. 8.13c). As a result, the voltage at the relay location increases to U_{Dmax} and could cause the measuring unit to pick up. To avoid this from happening, the relay setting is chosen to be safely above U_{Dmax}. The value of U_{Dmax} is given by

$$U_{Dmax} = \frac{I'_F \left(R_W + R_L \right)}{K_N} \tag{8.6}$$

where
I''_F = r.m.s. value of the fault current for a fault at the generator terminals,
K_N = rated c.t. ratio, R_W = c.t. secondary resistance, R_L = c.t. lead loop resistance

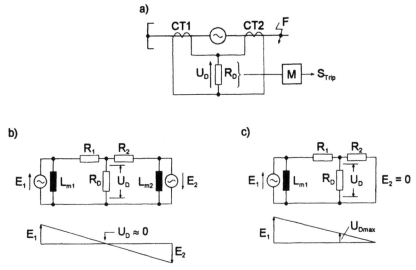

Fig. 8.13: Operating principle of a high-impedance differential protection for a generator
a) basic diagram, b) equivalent circuit and voltage distribution in normal
operation, c) equivalent circuit and voltage distribution for an external
phase fault with complete saturation of CT2

The operating times of both biased and high-impedance differential schemes
are of the order of 15 to 20 ms. The long DC time constants in the range of a few
hundred milliseconds characteristic of the fault currents of large generators [2.7]
can cause severe saturation of c.t's with closed iron cores and delay the operation
of the protection for faults in the zone of protection. This situation lead to the
development of protection schemes which are able to complete their measurement in the first few milliseconds of the fault, i.e. before any of the c.t's can start
to saturate. A scheme of this kind is shown in Fig. 8.14 [8.8, 8.28]. It is basically
a biased differential scheme in which in normal operation, the sum of the secondary currents I'_1 and I'_2 of the c.t's CT1 and CT2 flows through the resistor R_{rstr}
to generate the restraint voltage U_{rstr} and the difference current I_{op} flows through
the tripping c.t. TCT, the rectifier bridge TR and the resistor R_T to generate the
operating voltage U_{op}. Since, however, U_{rstr} is much greater than U_{op}, the protection restrains. The situation reverses for a fault in the zone of protection and U_{op}
becomes much greater than U_{rstr}. A current I_{R1} then flows through the relay AR1
which picks up and closes its contact K1. The rectified operating current flows
through the second relay AR_{op} which closes its contact K2. The operating time of
both AR1 and AR_{op} is about 1 ms and tripping takes place extremely quickly.
The tripping relay (not shown in Fig. 8.14) excited by the tripping signal S_{Trip}

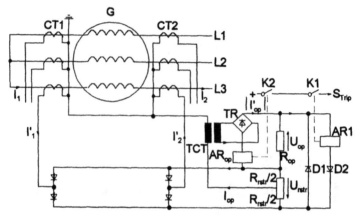

Fig. 8.14:   High-speed differential protection for large generators [8.8]

latches so that the tripping signal is maintained, should K1 and K2 open after the
1 ms as a result of c.t. saturation. By appropriate choice of settings and parameters, stable behaviour of the protection can also be achieved for an external fault
with c.t. saturation such that conventional c.t's with closed iron cores may be
used.

8.1.3.3 Time-overcurrent protection

Time-overcurrent protection frequently in combination with an undervoltage relay is used as back-up for a differential scheme in the case of generators and gen-
erator/transformer units with ratings of just a few tens of MVA. The protection
also performs the task, however, of isolating the generator from the power system, if external faults not detected by the differential protection are not cleared
by the corresponding transformer or line protection devices before the time set
for the back-up protection is reached. Figure 8.15 shows the principle of a two-
stage scheme in a simplified form. After the time set for the first stage (t_1), which
is time-graded with the other protection devices on the system, just the circuit-
breaker SCB connecting the generator to the rest of the system is tripped and
only if the overcurrent is still flowing after the time of the second stage ($t_2 > t_1$) is
the machine shut down completely.

Monitoring both current and voltage prevents false tripping of the back-up
protection due to an overload or during impulse excitation. The combination with
an undervoltage relay is essential where the supply for the generator excitation is
taken from the generator terminals [8.29], because the fault current contribution
by the generator decreases rapidly as the generator voltage and thus also the
excitation reduce for a fault close to the generator terminals [8.8, 8.30]. For this

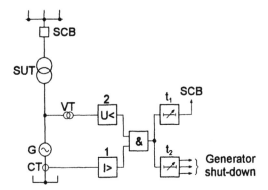

Fig. 8.15: Principle of time-overcurrent back-up protection with undervoltage supervision for a generator
1 = overcurrent relay, 2 = undervoltage relay

case, the combination of the two relays is arranged such that the undervoltage
relay enables tripping to take place although the overcurrent relay has reset
[8.11].

8.1.3.4 Distance protection

To overcome the relatively long operating times of the time-overcurrent back-up
protection described in the previous Section, distance relays are almost invariably
used for the back-up protection of large three-phase generators. Since large generators always operate in conjunction with a step-up transformer, the zone cov-

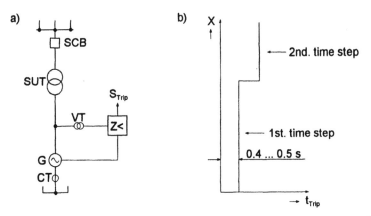

Fig. 8.16: Mode of connection a) and time-step characteristic b) of a two stage distance relay used as back-up protection for a large generator

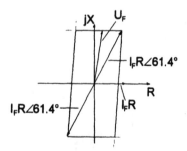

Fig. 8.17:   Operating characteristic of a distance relay used as back-up protection for a generator/transformer unit [8.32]

ered by the distance relay includes a portion of the step-up transformer, the cable between generator and the transformers (step-up and unit transformers) and a part of the unit transformer and the neighbouring high-tension transmission feeders leaving the associated busbars. The mode of connection and the characteristic of a distance relay used for this purpose are shown in Fig. 8.16.

The operating characteristic is usually a circle in the R/X plane with its center at the origin of the system of coordinates (see Fig. 6.21b) [8.8, 8.31], but may also be a parallelogram as shown in Fig. 8.17 [8.32]. The reach of the first zone is chosen such that about 70 % of the transformer winding (UT) is protected in addition to the generator. The second zone, on the other hand, should extend as far as possible into the power system without upsetting the grading of the various protection devices. As before, an undervoltage relay has to be added to sustain the starting function in the case of generators with excitation supplied from the generator terminals which cannot supply a continuous fault current [8.2].

8.1.4   Inter-turn protection

Inter-turn faults are breakdowns of insulation between turns of the same winding or between turns of parallel windings of the same phase. The probability of an inter-turn fault on medium to large generators having a single rod per slot is only slight, because the rods are well insulated from each other. Opinions therefore diverge over the need for inter-turn protection on machines with a single winding per phase and also on machines with parallel windings per phase, but a common star-point connection [8.11, 8.33]. Where parallel windings of a phase have individual star-point connections, inter-turn faults can be detected by a transverse differential protection or even a simple overcurrent relay between the two star-points.

The principle of a suitable transverse differential protection can be seen from Fig. 8.18a [8.34]. Only the center phase is shown in the diagram, but in practice

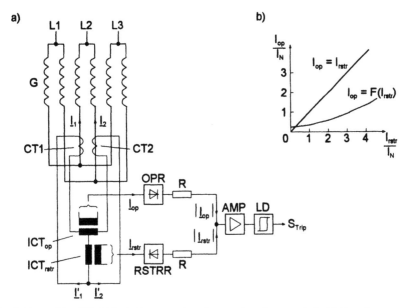

Fig. 8.18: Simplified block diagram a) and operating characteristic b) for a transverse differential protection scheme for detecting inter-turn faults [8.34]

the parallel windings of the three phases are supervised independently by identical circuits. The operating characteristic of the protection is similar to a longitudinal differential scheme and thus also has a restraining influence which increases with magnitude of through-current (Fig. 8.18b). Under normal load conditions, the restraint current I_{rstr} is the sum of the two winding currents I'_1 and I'_2 as transformed by the c.t's CT1 and CT2. Since for this case $I'_1 = I'_2$, after rectification

$$\left| I_{rstr} \right| = \left| I'_1 + I'_2 \right| = \left| 2I'_1 \right| \tag{8.7}$$

Under the same conditions, the operating current I_{op} is

$$\left| I_{op} \right| = \left| I'_1 - I'_2 \right| = 0 \tag{8.8}$$

and therefore mal-operation impossible.

The value of the circulating current caused by an inter-turn fault on one of the two windings and measured by the protection depends on the number of shorted turns. Assuming that the fault occurs when the generator is off-load, $I_1 = -I_2$ and therefore

$$\left| I_{rstr} \right| = \left| I'_1 - I'_2 \right| = 0 \tag{8.9}$$

and

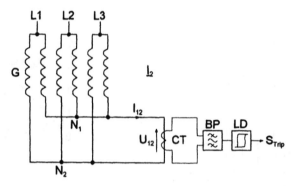

Fig. 8.19:   Simple inter-turn protection for generators with parallel windings

$$\left|\underline{I}_{op}\right| = \left|\underline{I}_1 + \underline{I}_2\right| = \left|2\underline{I}_1\right|$$ (8.10)

which causes the level detector LD to pick up and issue a tripping signal.

A second alternative for detecting inter-turn faults on parallel windings with separate star-point connections is shown in Fig. 8.19. The criterion monitored in this case is the voltage U_{12} between the star-points N_1 and N_2, respectively the balancing current I_{12}. Without a fault, the 50/60 Hz voltage $\underline{U}_{12} = 0$ and therefore $I_{12} = 0$. The voltage of the winding with an inter-turn fault reduces so that a corresponding difference voltage $\underline{U}_{12}$ and balancing current I_{12} result. This current is applied via the c.t. CT and the third harmonic filter F to the level detector LD. Tripping takes place if the current exceeds the setting of the protection of approximately 5 to 10 % of the generator rated current.

8.1.5   Rotor ground fault protection

Since the advent of excitation systems supplied by diodes mounted on the shaft of the machine or by thyristor circuits and the increase of the rotor-to-ground capacitance which took place at the same time as a consequence of the growing size of generators, the requirements a rotor ground fault protection scheme has to fulfil have become more complex. While previously the practice was for a duplicate ground fault protection scheme to only trip the machine when a second ground fault occurred and to simply signal single ground faults, it has become the rule today to trip single rotor ground faults as quickly as possible. This virtually excludes the possibility of duplicate ground faults which otherwise are extremely dangerous because of the thermal and mechanical effects they have on the rotor.

Protection schemes based on one of the following principles are currently in use to detect ground faults on the excitation circuit

-    measurement of conductivity using a 50/60 Hz voltage signal

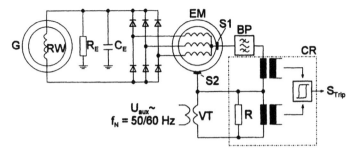

Fig. 8.20: Rotor ground fault protection based on measurement of conductivity for a synchronous generator with rectifiers mounted on the shaft [8.36]
RW = rotor winding, EM = exciter

- measurement of resistance using either a DC voltage or a low-frequency squarewave signal

The principle of the conductivity measurement is shown in Fig. 8.20 for a generator with rectifiers mounted on the shaft. The protection circuit comprises a bandpass filter BP, a conductivity relay CR and a v.t. VT and is connected to the star-point of the synchronous exciter EM via the slip-ring S1 and to the shaft of the machine via the slip-ring S2 [8.35, 8.36]. The bandpass filter BP is tuned to the fundamental of the auxiliary voltage U_{aux} (generally f_N = 50/60 Hz) which excludes all higher frequencies from the measurement. The equivalent impedance Z of the exciter circuit which in this example consists of the three phase windings of the exciter, the rectifiers and the rotor winding of the generator is given by the following relationships

- for a constant leakage resistance to ground R_E = const.

$$\left| \underline{Z} - \frac{R_E}{2} \right| = \frac{R_E}{2} \qquad (8.11)$$

- for a capacitive reactance $X_E = \dfrac{1}{\omega C_E}$ = const.

$$\left| \underline{Z} + \frac{X_E}{2} \right| = \frac{X_E}{2} \qquad (8.12)$$

where
C_E = ground capacitance of the excitation circuit

The family of orthogonal circles resulting from equations (8.11) and (8.12) are given in Fig. 8.21 [8.35]. It can be seen that applying the conductivity measurement according to equation (8.11), the leakage resistance R_E can be super-

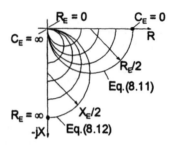

Fig. 8.21:   Family of orthogonal circles corresponding to equations (8.11) and (8.12) [8.35]

The family of orthogonal circles resulting from equations (8.11) and (8.12) are given in Fig. 8.21 [8.35]. It can be seen that applying the conductivity measurement according to equation (8.11), the leakage resistance R_E can be supervised uninfluenced by the ground capacitance C_E. From this follows that the operating characteristics of Fig. 8.22 provide optimum detection of rotor ground faults with circle 1 as an alarm stage and circle 2 as a tripping stage should the leakage resistance R_E fall to a dangerous level. Circle 2 is deliberately shifted towards the second and third quadrants to ensure reliable detection of solid ground faults (i.e. $R_E = 0$, see Fig. 8.21).

The conductivity principle described above has certain disadvantages, e.g. the quality of the measurement is influenced by the slip-ring resistances and, in the case of machines with rectifiers mounted on the shaft, by the behaviour of the excitation DC, and where the ground capacitance is large, the maximum value of the leakage resistance which can be measured is limited to about 10 kΩ. These disadvantages can be overcome by applying schemes which use either a DC or low-frequency squarewave voltage for measurement instead of one at power system frequency [8.37, 8.38]. Figure 8.23 shows the principle of a ground fault protection scheme for a generator with a separate thyristor supply which injects a

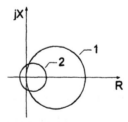

Fig. 8.22:   Operating characteristics of a rotor ground fault protection scheme based on the measurement of conductivity [8.35]
              1 = alarm stage,   2 = tripping stage

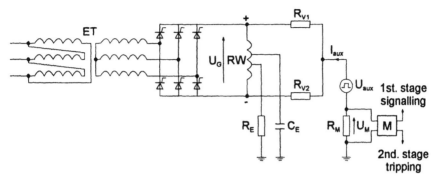

Fig. 8.23:   Measuring principle of a rotor ground fault scheme with squarewave injection signal [8.36]

squarewave at 1 Hz. The measuring circuit M is connected to the rotor winding via the high resistances R_{V1} and R_{V2}. The squarewave auxiliary voltage U_{aux} (Fig. 8.24) causes an auxiliary current to flow through the resistors R_M, R_{V1} and R_{V2} in parallel, R_E and the negligible resistance of the rotor winding R_W. The value of the auxiliary current I_{aux} is a measure for the insulation resistance R_E and is given by

$$I_{aux} = \frac{U_{aux}}{R_E + R_V + R_M} \qquad (8.13)$$

where
$$R_V = \frac{R_{V1} R_{V2}}{R_{V1} + R_{V2}}$$

Taking account of the fact that $R_M \ll R_V$, the measured voltage becomes

$$U_M \approx \frac{U_{aux} R_M}{R_E + R_V} \qquad (8.14)$$

The ground capacitance which is in parallel with R_E causes a sudden rise in the auxiliary current I_{aux} and therefore also of the measured voltage U_M at the beginning of every half-cycle of the auxiliary voltage U_{aux}. The duration of the corresponding damped oscillation depends on the value of the leakage resistance R_E. Regardless of the value of the ground capacitance C_E, the protection determines whether the steady-state value has been reached before measuring R_E. An example of the waveforms for two different leakage resistances is shown in Fig. 8.24. The scheme responds as soon as R_E falls below 80 kΩ and the first stage gives alarm. The second stage trips and shuts down the generator should the leakage resistance fall below 5 kΩ.

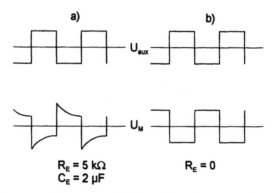

Fig. 8.24:   Auxiliary and measured voltages for different leakage resistances [8.36]

8.1.6   NPS protection

Unbalance loading of the three phases of a generator can result from a transmission line break, failure of a circuit-breaker pole or ground and phase-to-phase faults on the power system (see Section 2.4). The associated negative-sequence component in the stator current induces currents at double the power system frequency in the rotor, which cause an additional temperature rise mainly in the rotor retaining rings, the damping winding and the slot wedges. The degree of load

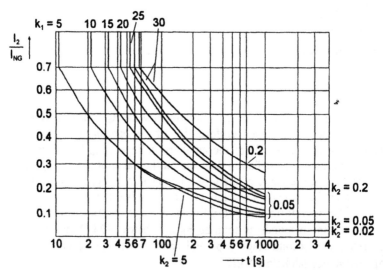

Fig. 8.25:   Inverse time operating characteristic of an NPS protection scheme [8.39]

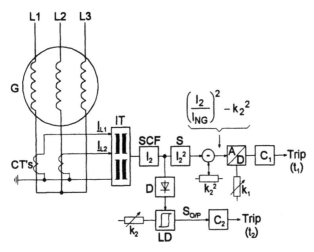

Fig. 8.26:   Simplified block diagram of an NPS protection scheme [8.39]

unbalance is indicated by the value of the negative-sequence current I_2 in relation to the generator rated current I_{NG}. The time during which it is permissible to operate a generator with an unbalanced load depends on its design and method of cooling (see equation (2.26)). The NPS protection therefore has an inverse characteristic. Typical operating characteristics given by the following relationship can be seen in Fig. 8.25.

$$t = \frac{k_1}{\left(\dfrac{I_2}{I_{NG}}\right)^2 - k_2^{\,2}} \qquad (8.15)$$

where

t	= operating time in s
I_2	= negative-sequence component of the current
I_{NG}	= generator rated current
k_1	= constant with a setting range of 5 to 30
k_2	= $0.775\dfrac{I_{2Z}}{I_{NG}}$, where I_{2Z} is the continuously permissible NPS current

The curves in Fig. 8.25 are obtained by inserting different values for the adjustable constants k_1 and k_2. The operating time becomes constant for values of the relative NPS current $I_2 > 0.7\ I_{NG}$, since such high values of NPS current can only occur during a fault and are then tripped by other protection devices. At the opposite end, the hyperbolic characteristic is limited to 1000 s, i.e. an unbalanced load condition which is not cleared earlier is tripped at the latest after 1000 s.

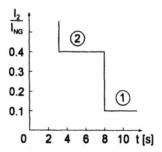

Fig. 8.27:   Two-step NPS protection characteristic
            1 = alarm,   2 = tripping

A simplified block diagram of an NPS protection scheme with the operating characteristics described above is given in Fig. 8.26. The input variables chosen in this case are the currents I_{L1} and I_{L2} which flow from the c.t. secondaries via the input transformers IT to the symmetrical component filter SCF. The latter extracts the NPS component and its relative value I_2/I_{NG} passes to both the squarer S and via the rectifier D to the level detector LD. The pick-up of LD is determined by the setting for k_2. The output signal O/P activates the electronic counter C_2 which trips after a time $t_2 = 1000$ s. The main counter C_1 which determines the equation (8.15) issues a tripping command after the time t_1 (see Fig. 8.25).

Apart from the NPS schemes with an inverse time operating characteristic, there are also schemes with a two-step inverse characteristic of the kind shown in Fig. 8.27, where one step is used for alarm (1) and the other (2) for tripping.

8.1.7   Loss of excitation protection

The loss of excitation protection guards against the consequences for the generator of a partial or complete failure of the excitation. The causes may be a short or open-circuit of the exciter, a failure of the automatic voltage regulator, an operator error under manual control or opening the excitation switch by mistake. Any of these circumstances endangers the stability of the generator which in most cases cannot then be prevented from falling out of synchronism. Local hot spots in stator and rotor are then the consequence. An open-circuit of the exciter has the added danger of the rotor insulation being damaged by overvoltages [8.40, 8.41].

A parameter which can be used to detect asynchronous operation resulting from loss of excitation is the ratio of the generator terminal voltage to the excitation current absorbed from the power system. From the point of view of the gen-

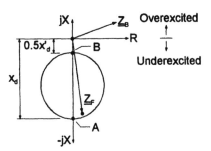

Fig. 8.28:   Operating characteristic of an ABB underreactance relay
Z_B = load impedance vector, Z_F = impedance vector after loss of excitation, X_d = synchronous reactance, x'_d = transient reactance

erator, drawing its excitation current from the power system is equivalent to
supplying a capacitive current. This means that the impedance vector Z_B at the
generator terminals in a loss of excitation condition shifts from the first to the
fourth quadrant of the R/X plane. Its value is determined initially by the synchronous reactance of the generator X_d, but after a time it approaches the transient
reactance x'_d. It follows from this that the condition can be detected by an underreactance relay having a circular characteristic with its center on the negative reactance axis (Fig. 8.28). This kind of characteristic can be achieved by the circuit
shown in Fig. 8.29. The input variables are the phase-to-phase voltage $U_{L2\text{-}L3}$ at
the generator terminals and the vectorial difference $I_{L2} - I_{L3}$. From the main v.t.
VT, the voltage $U_{L2\text{-}L3}$ passes to the interposing v.t. IVT, the tappings of which
enable the percentages of $aU_{L2\text{-}L3}$ and $bU_{L2\text{-}L3}$ to be set (b > a), corresponding to
the points A and B on the negative part of the ordinate in Fig. 8.28. The current
difference $I_{L2} - I_{L3} = I_{L2\text{-}L3}$ flows through the choke CH and produces the voltage
drop U_D. The generator current is purely capacitive if the excitation fails (Fig.
8.30a), i.e. the current $I_{L2} - I_{L3}$ leads the voltage $U_{L2\text{-}L3}$ by 90°. On the other hand,
the voltage U_D produced by the current $I_{L2} - I_{L3}$ leads the current by 90°. Thus a
phase-angle of 180° exists between U_D and $U_{L2\text{-}L3}$ (Fig. 8.30b). The following two
voltages are formed

$$U_1 = aU_{L2-L3} - U_D \qquad\qquad (8.16)$$

$$U_2 = bU_{L2-L3} - U_D \qquad\qquad (8.17)$$

and the phase-angle between them supervised. The protection operates when the
voltages U_1 and U_2 are in opposition (α = 180°) and restrains when they are in
phase (α = 0°). The pick-up limit is α_L = 90°. Factors a and b are chosen such
that points A and B (Fig. 8.28) fulfil the condition $bU_{L2\text{-}L3} > U_D > aU_{L2\text{-}L3}$.

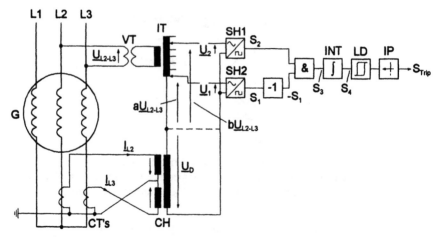

Fig. 8.29:   Block diagram of an ABB loss of excitation scheme using an underreactance relay

The waveforms of the electronic processing circuits of the protection scheme in Fig. 8.29 are shown in Fig. 8.31. It can be seen from the Figure that the limit angle α_L is supervised by determining the coincidence S_3 between the negative half-cycle of the voltage U_1 and the positive half-cycle of U_2. Since in this example $\alpha > \alpha_L$, the tripping signal S_{Trip} is issued and the generator would be tripped typically after a delay of 1 to 2 seconds.

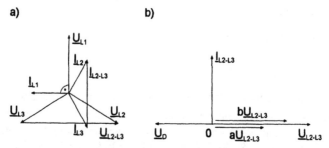

Fig. 8.30:   Current and voltage vectors for loss of excitation

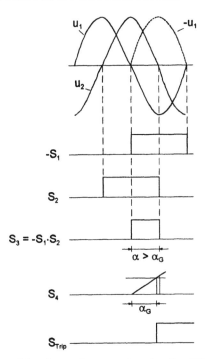

Fig. 8.31: Waveforms of the electronic processing circuits of the protection scheme in Fig. 8.29

8.1.8   Pole-slipping protection

The purpose of the protection dealt with in the last Section is to detect a loss of synchronism due to a failure of the excitation; the pole slipping protection also detects a loss of synchronism, but with the excitation intact. This condition can arise after a long power system fault or when a tie line between two systems is opened. This kind of loss of synchronism or out-of-step condition is accompanied by oscillations of real and apparent power in the system which on the one hand can threaten power system stability and on the other subject the slipping machine to severe mechanical stress. The unit's auxiliaries, especially any synchronous motors, are also at risk, because of the fluctuations of voltage and frequency which also occur.

The parameter supervised to detect generator pole slipping and out-of-step is the behaviour of the impedance vector measured at the generator terminals [8.18, 8.42, 8.43].

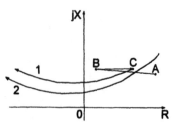

Fig. 8.32: Behaviour of the impedance vector at the generator terminals during out-of-step conditions following the tripping of a power system fault [8.18]
A = normal operation, B = beginning of fault, C = fault tripped
1 = first slip of the impedance vector, 2 = second slip of the impedance vectors

The typical behaviour of the impedance vector during the pole slipping which occurs after a power system fault can be seen from Fig. 8.32. The tip of the vector during normal operation is indicated by point A. At fault incidence it changes from A to B. When the fault is tripped, it jumps initially from B to C and then follows curve 1 for the first pole slip and curve 2 for the second. This behaviour of the impedance vector can be detected by a relay having the operating characteristic shown in Fig. 8.33. The characteristic comprises three parts, the lens (1), the straight line bisecting the lens (2) and the reactance line (3). In order to count as a pole slip, the impedance must enter the operating area at point M and leave it on the opposite side of the lens (point N). For this to take place, the center of the power swing must lie within the generator/transformer unit zone, in which case it is usual to trip the unit after the first slip. If the center of the power swing is lo-

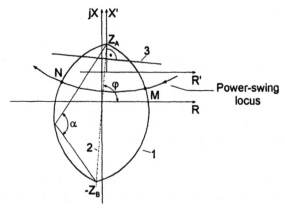

Fig. 8.33: Lenticular operating characteristic of a pole slipping protection relay [8.43]
Z_A, Z_B = power system and generator impedances

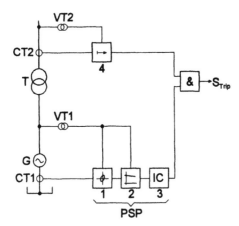

Fig. 8.34: Block diagram of a pole slipping protection scheme [8.43]

cated in the power system, i.e. above the reactance line (3), tripping is only initiated after a prescribed number of slips.

The block diagram of a pole slipping protection (PSP) is given in Fig. 8.34 and comprises an impedance measuring unit (1) with the lenticular characteristic, a reactance unit (2) to limit the reach in the direction of the power system and a impulse counter (3) which counts the slips and enables tripping when the set number is reached. The directional unit (4) connected on the HV side of the step-up transformer serves the same purpose as the reactance unit (2) of the pole slipping protection and provides for redundant operation of the protection.

Characteristics other than the one shown in Fig. 8.33 are also used for detecting pole slipping. An example is the parallelogram characteristic of Fig. 8.35 [8.45].

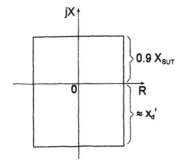

Fig. 8.35: Parallelogram characteristic of a pole slipping protection scheme [8.45]
X_{SUT} = reactance of the step-up transformer

8.1.9 Stator and rotor overload protection

Thermal overload of the stator winding can be caused by excessive reactive power load, i.e. operating at a low power factor, a fault in the cooling plant and failure of the voltage regulator when the machine is overexcited. The last two causes result in a temperature rise in the rotor winding as well. The rotor is also subjected to brief overloads due to voltage regulating operations when the machine is operating at full load.

It is clear that thermal overloads can only be tolerated for a short time. This applies especially to large base load generators with direct gas or water cooling which have only a short thermal time constant. Regulations exist [8.48] which define permissible overload currents for stator and rotor windings in relation to overload duration and form the basis for the operating characteristics for the corresponding overload protection. Typical protection characteristics are given in Fig. 8.36 for the rotor (1) and the stator (2) [8.43]. From these it can be seen that the protection only picks up when the permissible continuous rating (1.04 I_N) is exceeded. If the load current lies between this limit current and the lowest point of the curve, the operating time is no longer inversely proportional to the current and is constant at 300 s. Similarly, a fixed operating time of 10 s applies for currents higher than the range defined in the regulations.

Different operating principles are employed for thermal overload protection devices. The version shown in Fig. 8.37 is especially suitable for rotor windings [8.49]. The winding temperature is determined from two temperatures, the cooling medium temperature ϑ_c measured by a Pt 100 resistance thermometer and the temperature rise ϑ_{tr} of the winding modelled by the protection. A tripping signal S_{Trip} is issued when the corresponding criteria are fulfilled by both channels, i.e.

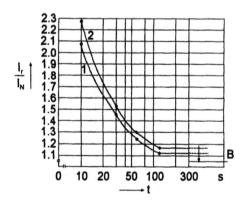

Fig. 8.36: Operating characteristics of the thermal overload protection devices for the rotors (1) and stators (2) of large generators [8.43].

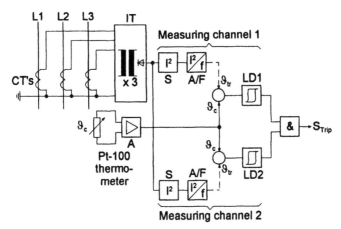

Fig. 8.37:   Simplified block diagram of a thermal overload protection scheme with a
             dual channel thermal image of the generator [8.49]
             IT = input transformer,   S   = squarer,   A/F = analogue/frequency
             converter,   A = amplifier, ϑ_{tr} = winding temperature rise,
             ϑ_c = coolant temperature

when both limit detectors LD1 and LD2 pick up. The limit detectors respond to
voltages proportional to the temperatures derived from digital counters (not
shown in Fig. 8.37).

8.1.10 Overvoltage protection

Inadmissibly high voltages at power system frequency are most commonly
caused by failure of the automatic voltage regulator or sudden load shedding.
The latter case is especially critical for hydro power plants, because the speed of
the machines rises rapidly and with it the generator voltage which can reach as
much as 200 %. The governors of both gas and steam turbo-alternator sets, on
the other hand, are designed to keep the transient overspeed within a few percent
when load is shed. The turbines are also equipped with emergency shut-off valves
which are capable of shutting off the steam or fuel supply within milliseconds
when a certain overspeed is reached. Overvoltages are primarily a hazard for the
generator insulation, but also cause over-fluxing of the step-up transformer, the
consequences of which are dealt with below (Section 8.2.4).

   The overvoltage protection for a generator typically comprises two time-
delayed overvoltage units connected to a phase-to-phase potential (Fig. 8.38).
The first unit (RU1) has a relatively low setting of about 1.1 U_N and acts to
weaken the excitation after its time delay has expired. The timer setting is determined by the time the voltage regulator needs to bring the voltage back down

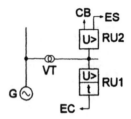

Fig. 8.38:   Block diagram of a two-stage overvoltage protection for a generator
             RU1, RU2 = overvoltage units,   CB = circuit-breaker,   ES = exciter
             switch,   EC = exciter controller (field weakening)

to its rated value after suddenly shedding load [8.2]. The second overvoltage unit
(RU2) has a higher setting of about 1.3 to 1.4 U_N and shuts the machine down
without delay.

8.1.11 Underfrequency protection

A reduction of power system frequency can be caused either by a deficit of real
power production in relation to power system load or a turbine governor fault.
The negative aspects of a reduction of power system frequency include

- damage to steam turbine blades due to vibration
- reduced capacity of the auxiliaries (pumps, fans, coal crushers etc.) and
  therefore of the whole power plant
- excessive temperature rise due to increased iron losses combined with reduced
  cooling efficiency
- over-fluxing of the step-up transformer

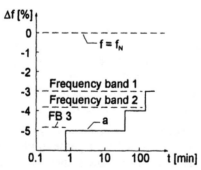

Fig. 8.39:   Permissible accumulated operating times for large steam turbo-alternator
             sets for different reductions of frequency [8.50]

As can be seen from Fig. 8.39, the time during which a turbo-alternator may be operated below rated frequency is limited [8.50]. The forbidden range lies below the stepped characteristic **a**.

The underfrequency protection can have several stages, but usually has two, the first one being set to about 47.5 or 57 Hz for 50 and 60 Hz systems respectively [8.2] and acting undelayed to isolate the turbo-alternator set from the power system. The second stage shuts the unit down after a few seconds should the frequency not recover in response to operation of the first stage.

Modern frequency relays measure the frequency digitally in relation to a quartz time base. Further details of digital frequency measurement are to be found in Section 12.6.

8.1.12 Power plant isolation protection

When a three-phase fault occurs close to a large turbo-alternator set, the set is relieved of its real power load, i.e. it accelerates and the rotor adopts a leading phase-angle in relation to what it was before [8.51, 8.52, 8.53]. If the fault is then tripped and the power system voltage once again applied to the generator either via other lines or high-speed auto-reclosure of the faulted line, there is generally a phase and amplitude difference between power system and generator voltages such that surges of torque are produced which for a large generator can be as much as 50 % higher than the stress of a solid fault at the terminals of the machine starting from a normal load condition. Apart from the electrodynamic stresses in the stator winding head region, the shaft train between the turbine and the generator is subjected to extreme torsional stress which reduces the life of the turbo-alternator set. The severity of the stress depends on both the magnitude of the negative step-change of real power at fault incidence and the duration of the fault.

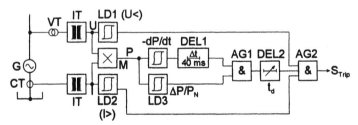

Fig. 8.40: Simplified block diagram of a power plant isolation protection scheme [8.52]

M = multiplier

The decisive criterion for determining whether a power plant has to be isolated from the power system is a negative step-change of real power $-\Delta P/P_N$ (P_N
= rated real power), however, overcurrent and undervoltage criteria are also included to ensure that isolation of the power plant only takes place, providing the
fault is not tripped by the corresponding protection device within the prescribed
time [8.52]. Figure 8.40 shows the block diagram of a power plant isolation
protection scheme, Fig. 8.41 the curves of the monitored variables and Fig. 8.42
the timing sequence when isolating a plant.

Not shown in Fig. 8.40 are the measuring units for the currents and voltages
of the other two phases. The corresponding variables are applied to the multiplier
(M) which also includes the processing functions to derive the three-phase real
power P. The occurrence of a three-phase fault close to the power plant causes a
negative step-change of real power from it normal load value P_L to the fault
value P_F (see Fig. 8.41c). The change in power is detected by a differential unit -
dP/dt which passes a signal to the AND gate AG1 after the fixed delay of 40 ms
set on DEL1. An output signal from AG1 is enabled by the level detector LD3
when P_F falls below its setting for $-\Delta P/P_N$. This signal enables the corresponding
input of the AND gate AG2 after the delay t_d of DEL2. Providing the three-
phase fault has not been tripped in the meantime, both the level detectors LD1

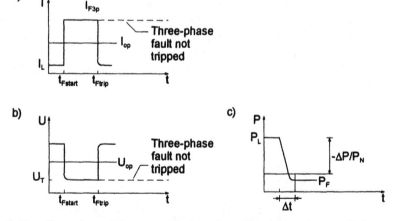

Fig. 8.41: Excursions of the current a), the voltage b) and the real power c) for a
three-phase fault close to a power plant [8.52]
I_L = generator load current, U_L = load voltage, P_L = normal real
power load, I_{F3p} = three-phase fault current, U_T = voltage at the
generator terminals for a three-phase fault, I_{op}, U_{op}, $-\Delta P/P_N$ = settings
for the monitored variables, t_{Fstart} = start of three-phase fault, t_{Ftrip} =
fault tripped by protection, Δt = set enabling time

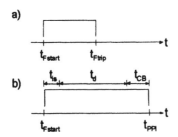

Fig. 8.42:   Timing sequence for the power plant isolation protection [8.52]
   a) correct tripping of the three-phase fault by the corresponding protection
      device
   b) three-phase fault persists and the power plant is isolated from the power
      system
   t_{Fstart} = start of fault, t_{Ftrip} = fault tripped by protection, t_{is} = operating time of the power plant isolation protection, t_d = delay setting
   for operation of the power plant isolation protection, t_{CB} = circuit-
   breaker operating time, t_{PPI} = power plant isolating time

(U<) and LD2 (I>) are excited (dashed lines in Fig. 8.41a and b), i.e. all three
inputs of the AND gate AG2 are enabled and the generator/transformer unit circuit-breaker is tripped. Tripping by the power plant isolation protection takes
place after the time t_{PPI} (see Fig. 8.42b).

   To ensure that the fault is detected, the undervoltage units LD monitor phase-
to-phase voltages. The operating voltage U_{op} is set to about 30 % of the generator rated voltage. The setting range for $-\Delta P/P_N$ is normally 60 to 76 % [8.2].

8.1.13 Reverse power protection

The reverse power protection protects the prime-mover from the effects of generator motoring. The generator motors when the supply of energy (steam, water
or gas) to the turbine fails. In the case of a steam turbine, the danger is overheating of the turbine blades as a consequence of having to compress the residual
steam; explosions can occur in the case of a diesel motor.

   The condition is detected by monitoring the direction of real power flow. A
very sensitive relay is needed, however, to detect the low levels of real power
consumed from the power system to cover the mechanical losses. For example,
the setting range for a steam turbine is 1 to 3 % and for gas and hydro units 3 to
5 %. Thus to improve the measuring accuracy of the protection, the current input
of the reverse power protection is connected in certain cases (large generator) to
the metering cores of the c.t's instead of the protection cores.

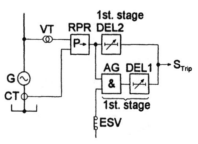

Fig. 8.43: Block diagram of a two-stage reverse power protection scheme
RPR = reverse power relay, ESV = auxiliary contact on the emergency stop valve

A reverse power scheme usually has two stages, the first stage isolating the unit from the power system in a time of 2 to 5 s after the emergency stop valves are closed and the second tripping the unit even if the emergency stop valves are open after a time of some tens of seconds. The block diagram of a two-stage reverse power protection scheme is shown in Fig. 8.43.

8.1.14 Guidelines for the choice of generator protection schemes

As explained in Section 8.1.1, a standard configuration of the protection for generators is impossible for many reasons. Of the many alternatives, only those combinations are discussed here which have achieved a degree of recognition as relatively standard solutions. Attention is also drawn once again to the fact that modern solid-state generator protection systems are all of modular design (see Chapter 5) [8.13, 8.14, 8.54].

A general statement can in as much be made that the number and complexity of the protection schemes installed is a function of the rated power of the unit to be protected, its significance for the power system of which it is part and its type (e.g. combined gas and steam turbine, motor generator in a pump storage plant etc.). Table 8.2 relates the types of protection needed to these factors.

Fig. 8.44 shows an example of the design of the generator protection for an industrial unit with a rating of 35 MVA. Most of the schemes which monitor current are connected to 5P20 class c.t's, while the reverse power protection (2) and the stator overload protection (4) are connected to the high-accuracy Class 0.2 metering cores.

For large generators which invariably are always generator/transformer units, the individual protection schemes are divided into two completely independent, electrically isolated groups, each with its own auxiliary supply units. The choice

Table 8.2   Choice of generator protection schemes

Type of protection	Generator rating [MVA]				ANSI No.
	< 5	5...50	50...200	> 200	
95 % stator ground fault	•	•	•	•	59N
100 % stator ground fault			•	•	64
Biased differential		•	•	•	87G
Time-overcurrent	•	•			50/51
Distance			•	•	21
Inter-turn		• [4]	• [4]		59
Rotor ground fault		•	•	•	64R
NPS		•	•	•	46
Underexcitation		•	•	•	40
Pole slipping				•	40
Stator overload	•	•	•	•	49
Rotor overload				•	49R
Overvoltage	•	•	•	•	59
Undervoltage	• [1]	• [1]	• [1]	• [1]	21
Underfrequency		•	•	•	81
Power plant isolation				•	
Reverse power	• [2]	• [3]	• [3]	•	32

[1] Only required by pump storage units while motoring and operation as phase compensator.
[2] Only steam turbine and diesel prime-movers
[3] Not necessary for Pelton hydro units
[4] Only necessary where several rods belonging to the same phase are in the same slot.

of schemes in the two groups is made firstly to achieve redundancy of protection between the groups and secondly to permit testing of each group in turn while the generator is on load by testing devices, which today are part of the standard equipment belonging to a generator protection. An example of dividing a protection system into two redundant groups is shown in Fig. 8.45 for a large turbo-alternator set. Not shown is the protection for the step-up and unit transformers which is dealt with in detail together with assignment of the tripping signals of the individual protection schemes to the switchgear (tripping logic) in Section 8.2.

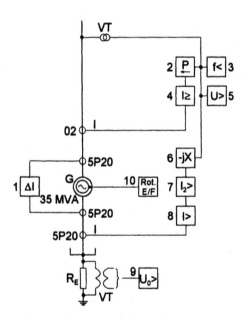

Fig. 8.44:   Generator protection configuration for an industrial unit with a rating of 35 MVA
1 = biased differential,   2 = reverse power,   3 = underfrequency,   4 = stator overload,   5 = overvoltage,   6 = undervoltage,   7 = NPS,   8 = time-overcurrent,   9 = 95 % stator ground fault,   10 = rotor ground fault,   RE = grounding resistor

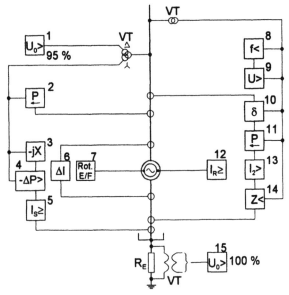

Fig. 8.45: Scope of protection and grouping of schemes for a large generator
1 = 95 % stator ground fault,   2 = reverse power No. 1,   3 = underexcitation,   4 = power plant isolation,   5 = stator overload,   6 = biased differential,   7 = rotor ground fault,   8 = underfrequency,   9 = overvoltage,   10 = pole slipping,   11 = reverse power No. 2,   12 = rotor
overload,   13 = NPS,   14 = distance,   15 = 100 % stator ground fault

8.2   Protection of generator/transformer units

8.2.1   The generator/transformer unit as protected unit

A considerable number of different configurations for protecting power plant
units and their auxiliaries has become established in order to satisfy the requirements at the different power plant ratings, the locations of the power plants in the
system and the boundary conditions prevailing in the particular system. It would
go beyond the scope of this book to explain and analyze all the possible circuits
with respect to quality of function. Nevertheless, it is important to be aware of
the basic considerations concerning protection design, protection zones and fast
discriminative tripping.

The procedure for selecting the phase fault protection for a genera-
tor/transformer unit is illustrated in Fig. 8.46. The zones of protection for the
various items of primary system plant must be chosen such that they firstly over-

lap while achieving the shortest possible tripping times and that secondly back-up protection with relatively short operating times is provided [8.9]. In this example, the generator/transformer unit comprises the generator (G), the step-up transformer (T), the unit transformer (UT), the excitation circuit including the transformer (ET), the generator circuit-breaker (GCB), the power system circuit-breaker (SCB) and the cables connecting these items of plant. Most important is to note that there are no unprotected gaps, a remark which applies generally throughout protection.

The specialized protection schemes for the individual items of primary plant shown in Fig. 8.46 have already been explained in previous Chapters and Sections. It is therefore only necessary to discuss at this juncture those protection devices and schemes which are specifically designed for generator/transformer units. It should also be noted that zones of certain protection schemes explained in connection with generator protection are extended to include other items of plant belonging to a generator/transformer unit, e.g. the stator ground fault pro-

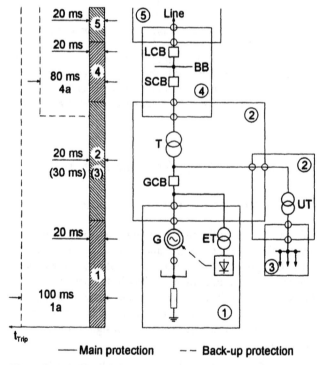

Fig. 8.46:   Phase fault protection zones for a generator/transformer unit
1 = generator zone,   2 = step-up and unit transformer zone,   3 = unit auxiliaries busbar zone,   4 = busbar zone,   5 = transmission line zone

tection which also detects ground faults between the generator terminals and the inputs of step-up and unit transformers. Similarly, distance and time-overcurrent relays measuring at the generator star-point reach into the transformer windings.

8.2.2 Biased differential schemes

The number of differential schemes installed and their respective zones depends mainly on the basic configuration of the generator/transformer unit concerned. A typical arrangement for a large unit can be seen in Fig. 8.47 which omits the interposing c.t's for simplicity. In this example, there are three independent differential schemes for detecting phase faults. Protection 1 detects faults on the generator stator windings, protection 2 on the step-up transformer and the cables between the c.t's near the generator terminals and the c.t's on the outgoing transmission line and protection 3 on the unit transformer and its connections. The configuration of the protection in Fig. 8.46 fulfils the requirements with regard to minimum tripping times for phase faults in the entire generator/transformer unit zone.

8.2.3 Distance relay as back-up protection on the HV side of the step-up transformer

A distance protection scheme was described in Section 8.1.3.4 for detecting phase faults between the windings of synchronous generators and partly in the

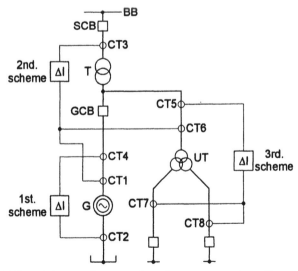

Fig. 8.47: Differential protection zones for a large generator/transformer unit

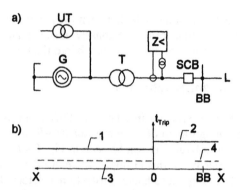

Fig. 8.48:   Connection a) and time-step characteristic b) for a distance relay connected
on the HV side of the step-up transformer
1 = forwards measurement,   2 = reverse direction,   3 = genera-
tor/transformer unit differential protection,   4 = first zone of the transmission line distance relay

step-up transformer. In order to maintain proper discrimination, a part of the
transformer windings can only be protected by this relay in the second zone (see
Fig. 8.16). Thus to ensure that faults in this part of the windings are also tripped
in the shortest possible time, a second distance relay is installed on the HV side
of the step-up transformer. Figure 8.48a shows how the relay is connected and
Fig. 8.48b the corresponding time-step diagram. As can be seen, the relay responds to faults both in the forwards (line 1) and reverse directions (line 2). Thus
fast back-up protection is provided not only for the differential protection devices
(line 3), but also for the busbar (BB) and the transmission line (line 4).

8.2.4   Overfluxing of the step-up transformer

The danger of overfluxing a step-up transformer (i.e. exceeding the rated flux of
the iron core) and entering saturation exists primarily when the transformer is
connected on the LV side to the generator, but not connected to the power system on the HV side [8.55, 8.56]. The initiating factor in these circumstances may
be an increase in voltage or a reduction of frequency caused by incorrect starting
up or tripping of the unit at full load accompanied by voltage regulator or speed
governor failure.

The additional iron losses in the saturated region heat up the iron core. The
flux also leaks into other parts of the transformer to produce eddy currents and
still more heat in the windings and structural parts.

A measure of the degree of overfluxing of the transformer is the ratio of the
saturation flux B_S to the rated flux B_N. According to regulations, the continu-

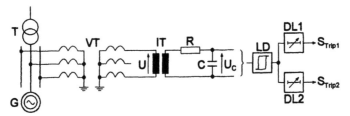

Fig. 8.49: Block diagram of an overfluxing protection scheme [8.55]
            R, C = voltage divider

ously permissible value of the ratio B_S/B_N is limited to 1.1. A suitable setting for
the overfluxing protection is therefore $B_S/B_N = 1.2$ to 1.4. The measuring principle of the protection is based on the following relationship between the flux B,
the frequency f and the voltage U induced in the transformer

$$B = \frac{1}{4.44na} \cdot \frac{U}{f}$$
(8.18)

where
n = number of turns,   a = the cross-sectional area of the iron core

From this, the ratio B_S/B_N becomes

$$\frac{B_S}{B_N} = \frac{U}{U_N} \cdot \frac{f_N}{f}$$
(8.19)

where
U_N = rated voltage, f_N = rated frequency

According to equation (8.19), the per unit values of voltage and current indicate the degree of overfluxing and it is these values which are used in practical
overfluxing relays.

Figure 8.49 shows an example of a practical overfluxing relay. The input
variable is the phase-to-phase voltage U measured by the v.t. VT which is cou-
pled by the input transformer IT to the potential divider R/C. The small portion
of the voltage across the condenser C is applied to the level detector LD. Since
R is much greater than $1/\omega C$, $U_C \approx U/f$. The level detector LD picks up when the
setting in per unit is exceeded. Typically the level detector excites two timers,
timer 1 (DL1) which reduces the excitation after 1 s (signal S_{Trip1}) and timer 2
(DL2) which shuts the unit down after 10 to 20 s, if the overfluxing condition
persists (signal S_{Trip2}).

8.2.5   Miscellaneous generator/transformer unit protection devices

Apart from the protection devices for power transformers (Chapter 7), generators (Section 8.1) and generator/transformer units (Sections 8.2.2 to 8.2.4), further protection devices may be included according to the type of prime-mover (gas or hydro turbines etc.) or operating mode (motor/generator in a pump storage plant), e.g. under and over-speed protection against excessive load in pump operation, or undervoltage protection against failure of the supply system when operating as a motor or phase compensator.

One of the special protection schemes used on gas turbine and pump/storage units is the DC protection of the start-up converter [8.54]. A typical arrangement of a start-up converter (SUC) and DC protection is shown in Fig. 8.50. In the event of a ground fault F in the DC link circuit, the star-points of both the start-up transformer SU and the generator are raised with respect to ground by a voltage comprising a DC and a superimposed AC at three times the fundamental frequency of the power supply. The value of the off-set voltage is determined by the response of the converter controller and the distribution of the ground capacitances and can reach as much as 67 % of the r.m.s. value of the start-up transformer phase-to-phase voltage. This voltage can prove a hazard for any items of plant connected to ground such as v.t's and grounding transformers.

A ground fault in the DC circuit is detected by monitoring the voltage drop it causes across a shunt. A DC/DC converter reduces the signal level for measurement and provides electrical insulation between primary system and measuring circuit. The signal level is then measured by the DC relay DCR. A flashover fuse is connected in parallel with the shunt which instantly grounds the star-point of the v.t's in the event of an open-circuit of the shunt or its leads.

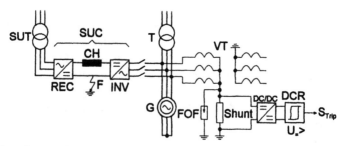

Fig. 8.50:   General diagram of a generator/transformer unit with a start-up converter and DC protection scheme [8.54]
REC = rectifier,   INV = inverter

8.2.6   Design of a protection system for a generator/transformer unit

As mentioned previously, no general rules can be given for the design and scope of the protection for a generator/transformer unit. Therefore just two alternatives for two different kinds of plant are described below as examples.

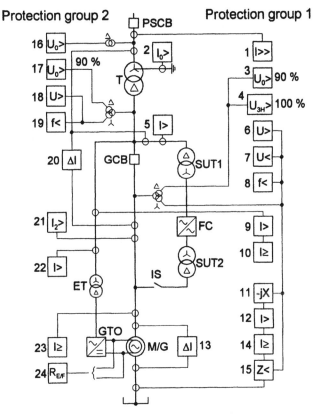

Fig. 8.51: Scope of protection and grouping of schemes for a motor/generator/transformer unit in a pump storage plant [8.10]
M/G = motor/generator,   T = step-up transformer,   ET = excitation transformer,   SUT = start-up transformer,   FC = frequency converter, GTO = GTO thyristor excitation,   GCB = generator circuit-breaker, PSCB = power system circuit-breaker,   IS = isolator,   I>> = overcurrent,   I≥ = overload,   I> = time-overcurrent,   U> = overvoltage, U< = undervoltage,   f< = underfrequency (underspeed),   U₀ = ground fault,   ΔI = differential,   Z< = distance,   I₂> = NPS,   -jX = underexcitation,   R_{E/F} = rotor ground fault

Figure 8.51 shows the general overview of the protection system for a pump storage unit with a rating of 60 MVA. The machine can operated as a generator, a motor or in a phase compensation mode. The start-up plant for the motor mode comprises the transformers SU1 and SU2 and the frequency converter FC. The excitation for the machine is supplied by the transformer ET and the thyristor controller GTO. Note that here too the individual schemes are divided into two independent redundant groups for greater reliability and to facilitate testing of the protection while the unit is in operation. The two groups are composed such that each is able to detect all the kinds of faults. Compared to the protection for a steam-driven set, additional schemes are included for the start-up and steady-state operating phases in the motoring mode, e.g. overcurrent devices 1 and 23 and the undervoltage device 7.

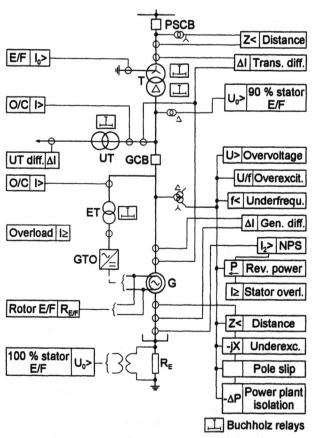

Fig. 8.52:   Kinds of protection typically installed on a high-power turbo-alternator set
             ($P_N \geq 600$ MW)

Some minor or non-electrical kinds of protection have been omitted from Fig. 8.51, e.g. the protection devices for the frequency converter and the Buchholz relay for the step-up transformer.

The protection system for a high-power turbo-alternator set is shown in Fig. 8.52. Although to simplify the layout, the division into two groups is not visible in the diagram, the schemes are indeed divided into two groups according to the considerations discussed above (see Fig. 8.45).

8.2.7 Allocation of the protection tripping signals to the switchgear and turbine valves (tripping logic)

The tripping signals of the individual protection schemes must go to the appropriate switchgear and turbine valves as dictated by the design of the generator/transformer unit, its operating mode (generator, motor or compensator) and the kind of fault. A typical arrangement is shown in Fig. 8.53 for a high-power turbo-alternator set equipped with a generator circuit-breaker GCB. The example assumes that the protection system PS is required to shut down the unit completely and give alarm.

A modern, modular, solid-state protection system includes a programmable tripping logic which can be programmed, for example, by inserting diode plugs in a matrix distributor (see Fig. 5.18 in Section 5.3.5). Adhering to the policy of

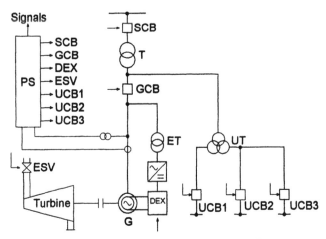

Fig. 8.53:   Allocation of tripping signals for a generator/transformer unit
PS = protection system,   SCB = power system circuit-breaker,   GCB = generator circuit-breaker,   DEX = de-excitation switch,   ESV = turbine emergency stop valve,   UCB = unit auxiliaries circuit-breakers

dividing the protection into two redundant groups for greater reliability, there are also two completely identical, but electrically insulated distribution matrices for protection group 1 and 2 respectively (also designated "left" and "right"). The double diode plugs can be inserted without restriction to connect the tripping signals coming from the protection devices to the desired tripping circuits.

An example of the allocation of tripping signals of the two protection groups for a high-power turbo-alternator set is given in Fig. 8.54. Note that the tripping logic also allocates the tripping signals from external protection devices such as Buchholz relays.

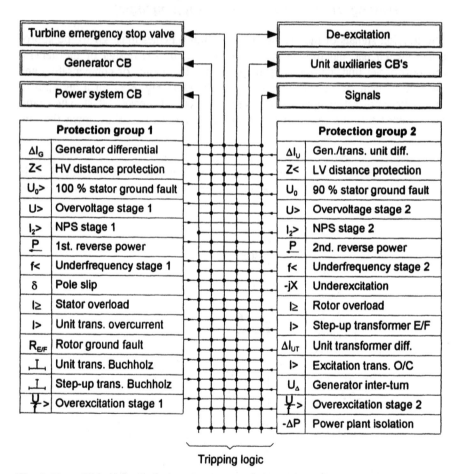

Fig. 8.54: Allocation of tripping signals to the switchgear of a large generator/transformer unit via a distributor matrix tripping logic

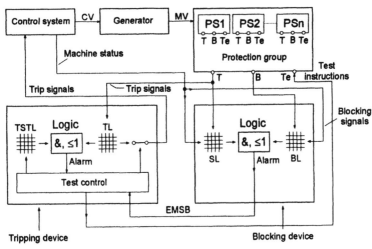

Fig. 8.55: Schematic of tripping and blocking logics for a solid-state generator protection system [8.57]
PS = protection system,   T = trip,   B = block,   Te = test facility,
CV = control variable,   MV = measured variable,   TSTL = test logic,
TL = tripping logic,   SL = supervision logic,   BL = blocking logic,
EMSB = emergency switch-back to the test mode should blocking fail

In some cases it is necessary to block protection devices to prevent false tripping in certain operating modes. This applies, for example, to current and voltage measuring devices during the start-up phase of gas turbines and pump storage units [8.54], because the c.t's and v.t's can saturate temporarily at the very low frequencies involved. Protection devices designed for rated frequency may also be adversely influenced. To provide for such cases, there is also a blocking logic in addition to the tripping logic [8.14]. A schematic of tripping and blocking logics is shown in Fig. 8.55 [8.57], from which it can be seen that the machine status is applied to the horizontal bars of the supervision and blocking logics (matrices). According to the locations of the diode plugs, the blocking commands are relayed via the vertical bars of the blocking matrix to selected protection devices.

8.3   Protection of HV three-phase motors

8.3.1   Kinds of faults and protection schemes

Internal and external faults and operating conditions threaten HV motors. Internal faults include faults between windings, inter-turn faults and ground faults and external faults and operating conditions include undervoltage, overload during starting and steady-state operation, asymmetric loading due to phase imbalance, phase and ground faults on the motor feeder and asynchronous operation of synchronous motors.

Protection against most of these faults is afforded by protection devices which have already been dealt with in connection with other items of plant [8.58, 8.59], however, because of the specialities of induction motors and synchronous motors, new criteria have to be monitored in some cases. Table 8.3 lists the kinds of faults and the protection devices used to detect them. Examination of the table discloses that an inter-turn fault is the only kind of fault for which no protection is included. The reason for this is that it is extremely difficult to detect, since it causes no positively identifiable change in the external electrical variables. The detection of an inter-turn fault is therefore left to the ground fault protection in the assumption that it will develop into a stator ground fault. The following sections briefly explain the various protection schemes.

Table 8.3   Kinds of faults and protection devices for HV three-phase motors

Kinds of faults		Protection devices	ANSI No.
Phase faults between the stator windings or on the motor feeder		Overcurrent or differential	50/51
Phase faults on a stator winding or the motor feeder		Overcurrent or directional	87
Stator inter-turn fault		-	
Overload	Running up	Overload, start-up time supervision	49
	Operation	Overload	
Phase imbalance		NPS	46
Voltage dips		Undervoltage	27
Asynchronous operation of synchronous motors		Out-of-step	40

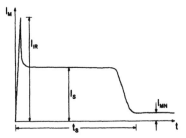

Fig. 8.56: Stator current (r.m.s.) when starting a squirrel-cage induction motor
I_{IR} = inrush current, I_S = starting current, I_{MN} = motor rated current

8.3.2 Phase fault protection

8.3.2.1 Overcurrent protection

Generally fast-operating two-phase overcurrent devices are used to detect phase faults between stator windings and on the motor feeder. The setting is chosen such that the protection does not trip incorrectly when the motor is switched on, but genuine stator phase faults are reliably detected. Fulfilling these two conditions for a squirrel-cage induction motor can be very difficult, mainly because of the current curve upon energizing the motor. This problem is illustrated in Fig. 8.56 which shows a current inrush as high as 15 times the rated current of the motor immediately after switching on [8.59]. A short time delay of 2 to 3 periods (33 to 50 ms at f_N = 50/60 Hz) is unavoidable to prevent this inrush current from causing false tripping. The delay permits the protection to be set just above the starting current I_S. Since this usually lies between 3 and 6 times I_{NM}, it is impossible to detect phase faults near the star-point of the motor. With such a high setting, it may also happen that the protection cannot pick up, if the available short-circuit power is low.

The starting current of slip-ring induction motors is appreciably lower and a lower setting for the protection is therefore permissible which extends its zone towards the star-point.

8.3.2.2 Differential protection

Biased differential relays are used to protect both synchronous and induction motors against phase faults (see Section 6.2). This requires, of course, two sets of c.t's, one at the beginning of the motor feeder, the other at the star-point. The basic setting I_g is not set higher than 0.5 I_{NM} which considerably improves the sensitivity of the protection compared with the overcurrent protection described above. Judicious choice of the restraint factor k_{rstr} (see Fig. 6.11) enables false tripping for phase faults on the supply system to be avoided.

8.3.3   Stator ground fault protection

The design of the ground fault protection for the stator windings and the motor feeder depends on the principle employed for grounding the power system. A ground fault current can easily be detected by a three-phase overcurrent scheme in a solidly or low-impedance grounded system, because the ground fault current is higher than the motor rated current (see Section 8.3.2.1). The ground fault protection in ungrounded or high-impedance grounded systems can be one of two types:

1. overcurrent ground fault protection
2. directional ground fault protection

The use of **overcurrent ground fault protection** is permissible, providing the capacitive ground fault current of the electrically connected power system is at least twice the setting on the overcurrent relay. The standard scheme comprises a sensitive overcurrent relay supplied by a core-balance c.t. at the beginning of the motor feeder (see Section 7.1.3) and responds to the zero-sequence component. The zone protected usually extends to 70 to 80 % of the stator windings, i.e. 20 to 30 % of the stator windings remain unprotected.

Where the zero-sequence component of the ground fault current is too low to ensure detection by the overcurrent scheme described above or the reach into the stator winding is too short, **directional ground fault protection** is used to monitor either the apparent power component (ungrounded system) or the real power component (system with Petersen coil) of the ground fault current (see Section 7.1.3). The zero-sequence component of the neutral (off-set) voltage obtained from a broken delta arrangement of the v.t's on the busbars is usually used as the reference quantity needed to determine direction.

Both the above schemes act in most cases to trip the motor being protected after a short delay.

8.3.4   Overload protection

8.3.4.1 Overload during starting

It can be seen from Fig. 8.56 that the starting current of an induction motor is very much higher than its load current, i.e. the thermal stress on the motor is also correspondingly higher during the starting period than in steady-state operation.

Modern high-power three-phase motors are generally designed such that over-loads must be kept within closely prescribed limits and therefore the number of consecutive starts starting from cold is confined to three, respectively two when warm, [8.59] and a maximum duration must not be exceeded for each attempt. On the other hand, it can be necessary in certain circumstances to be able to start a motor when warm, although the maximum continuously permissible tempera-ture is briefly exceeded.

A distinction is made between overload protection devices which model the temperature rise of the rotor and those which model the temperature rise of the stator. In the former case, the protection monitors the start-up procedure in rela-tion to an accurately chosen overload characteristic. The latter case additionally supervises the starting time and is therefore also able to guard against motor stalling. The scheme comprises a definite time-overcurrent unit with a current setting lower than the motor starting current and a delay setting longer than the normal starting time t_S of the motor (see Fig. 8.56). This means that the overcur-rent measurement picks up when the motor is being started, but does not trip be-cause the time delay setting t_{DEL} is longer than t_S. The overcurrent unit resets for $t = t_S$. The stator current is always higher than the current setting (even for $t = t_{DEL}$) in the event of a fault, e.g. a stalled motor, and the motor is tripped.

8.3.4.2 Overload during steady-state operation

The usual causes of thermal overload during normal steady-state operation are mechanical damage to the motor or the machine it is driving, a supply voltage reduction or reduced coolant supply. In the first two cases, the overload can be detected by monitoring the stator current. The protection devices used for the purpose may have a partial or full memory function [8.59, 8.61]. The fundamen-tal difference between the two is that the normal load current is neglected by a partial memory function, while the full memory function takes it into account (see Section 2.5). The technical design of an overload protection device with full memory function, i.e. thermal image of the protected unit, is similar to that shown in Fig. 8.37 which also takes account of the ambient temperature and thus permits optimum matching of the operating characteristic to the motor.

8.3.5   NPS protection

Asymmetrical load in the case of HV motors is mainly the result of an imbalance of the three-phase supply voltages. Other causes are only of a transient nature (DC component of the stator current during starting) or are only possible in ex-ceptional circumstances, e.g. open-circuit conductor in the supply system, the motor feeder or the stator windings.

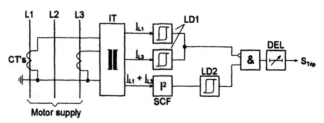

Fig. 8.57:   Block diagram of an NPS protection for an HV three-phase motors
             SCF = symmetrical component filter

In view of the fact that the negative-sequence reactance of a three-phase motor is 5 to 7 times smaller than the positive-sequence reactance, even slight asymmetries of the three-phase supply cause relatively high negative-sequence components in the stator currents. For example, for an induction motor with a ratio I_S/I_{MN} = 5 (I_S = starting current, I_{MN} = motor rated current), a negative-sequence voltage component of 3 % corresponds to a negative-sequence current component of 15 %. The latter induces a three-phase field in the rotor which rotates in the opposite direction to the rotor's mechanical direction and causes an additional temperature rise. Complete loss of a phase results in an additional temperature rise of the stator as well.

The NPS protection for HV three-phase motors is normally only equipped with a single stage with a setting in the range 10 to 20 % of the motor rated current. Upon picking-up, the protection trips the motor after a delay of a few seconds [8.60]. The negative-sequence component in the stator current is detected by a symmetrical component filter connected to the main c.t's of two of the phases (Fig. 8.57). A signal proportional to the negative-sequence component is monitored by the level detector LD2. The two other level detectors LD1 supervise the values of the phase currents I_{L1} and I_{L3}. If these exceed 5 I_{NM}, the AND gate is blocked to prevent the NPS protection from mal-operating while the motor is starting.

It should be remembered, of course, that a phase-to-phase stator fault produces a negative-sequence fault current I_2, so that where only an overcurrent relay is installed to detect phase faults, the NPS protection also serves as a sensitive phase fault protection for phase-to-phase faults close to the star-point.

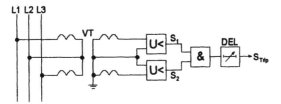

Fig. 8.58:   Block diagram of an undervoltage scheme for HV three-phase motors

8.3.6   Undervoltage protection

The undervoltage protection has the task of detecting inadmissible voltage reductions or complete voltage failure and trip a motor or a series of motors. The two main fields of application for this kind of scheme are

1. three-phase motors which for technical or safety reasons are not permitted to start on their own when the supply is restored after an interruption

2. to trip a group of less important drives to facilitate the self-starting of important induction motor drives when the supply is restored.

Synchronous motors belong to the first group which are likely to fall out-of-step if the supply voltage is too low.

The corresponding protection scheme comprises two or three voltage relays which initiate tripping via a timer. The pick-up voltage is set to 60 % of rated voltage and the timer to between 0.5 and 0.7 s. The protection is usually connected to the v.t's for the entire plant (metering bay). The v.t. secondary circuits must be themselves protected by m.c.b's in order to distinguish between voltage failure of the primary system and voltage failure of the v.t. secondary circuit.

The block diagram of an undervoltage scheme is shown in Fig. 8.58. In this example, mal-operation due to failure of a v.t. secondary circuit is avoided by logically linking the signals S_1 and S_2 in the AND gate.

8.3.7   Out-of-step protection for synchronous motors

A synchronous motor can be made to fall out-of-step and then run asynchronously by a voltage reduction, mechanical overload, power swings and excitation circuit faults. The consequences are thermal stress of the rotor, mechanical shock to the shaft and couplings and, in the case of rectifier excitation, dangerous overvoltages in the field winding [8.58].

There are different methods of detecting loss of synchronism. One relies on the fact that the motor draws its reactive power from the power system when running asynchronously. Thus the condition can be detected by installing a sensitive reverse power relay with a characteristic angle of 90° which is connected to the phase currents and voltages at the beginning of the motor feeder. A few seconds after picking up, the protection trips the circuit-breaker and de-excites the motor.

An alternative solution detects the pulsating of the stator current which asynchronous operation causes [8.21].

8.3.8 A complete protection system for HV three-phase motors

From the discussion in Sections 8.3.1 to 8.3.7, it follows that the protection system for an HV three-phase motor can be composed in different ways depending on the motor's type, rating and application. This procedure, however, has been appreciably simplified since the advent of modularity in protection which permits the number and types of protection devices to be combined at will. Figure 8.59 shows two examples for HV three-phase motors of medium and high power (a few MW). Figure 8.59a concerns an induction motor with an ungrounded starpoint. This is decisive for the choice of the overcurrent protection (2) to detect phase faults. It was also assumed that the capacitive ground fault current is sufficiently high to permit the use of an overcurrent unit to detect ground faults.

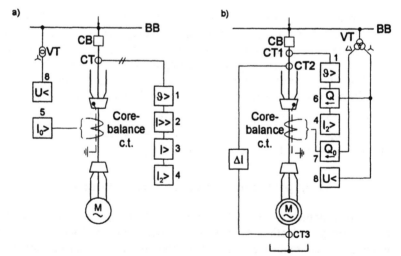

Fig. 8.59: Integrated protection systems for a) a medium power HV induction motor and b) an HV synchronous motor
1 = steady-state overload,   2 = phase fault overcurrent,   3 = maximum starting time,   4 = NPS,   5 = ground fault overcurrent,   6 = out-of-step,   7 = directional ground fault,   8 = undervoltage,   9 = differential

The protection system of Fig. 8.59b is suitable for a synchronous motor with a rating of a few MW. Phase fault protection in this case is afforded by a differential unit (9). Because of the low ground fault level, the overcurrent ground fault protection (5) has to be replaced by the directional ground fault protection (7).

The higher the rating and value of a synchronous motor, the closer its protection system approaches that of a synchronous generator (see Section 8.1).

Part C
Digital Protection

9

Computer-Based Protection and Control

9.1 Introduction

Digital computers, initially used in electrical power systems for off-line calculations, statistics etc., and then for planning and forecasting, soon found their way into load dispatching centers and on-line applications. These increasingly involved time-critical processing and safety control tasks and proved the reliability of industrial digital engineering techniques in the power system environment. The appearance of microprocessors enabled ever simpler tasks to be performed at distributed locations and their use for protection and control was logically the next step. Thus memory and computing capacity is now available at "feeder" level which enabled the processing of additional parameters and also new parameters that could not be processed previously. The easy communication with other computers and data bases provides access to data on other command levels and other areas of control. Functions have also been transferred from the hardware to the software level, thus making possible system adaptation and expansion by software configuration.

A major advantage of digital techniques is their ability to continuously self-monitor important circuits and functions and ensure the uninterrupted availability of the devices.

All these innovations basically change the procedures for engineering, erection, commissioning, maintenance and also subsequent system expansion and, if performed properly, also reduce the corresponding costs. They do, however, have an influence on the organizational structures of the users and sub-suppliers.

It is therefore an operational question to what extent control, measurement, protection, metering and plant diagnosis functions should be integrated. The multiple utilization of process signals (e.g. c.t. currents), the ability to logically relate data from different sources and many other possibilities would support such a development. It is absolutely essential, on the other hand, that at each step on the way to integration, vital functions such as protection are not adversely influenced by the links with other functions and also that the integrated system as a whole conforms to the strictest requirements of the component functions, which in our case are those applicable to protection.

9.2   Digital protection and control system structures

9.2.1   General

Because of their geographic size, the concentration of energy, the fluctuations of load and the safety requirements, the management of large electricity supply systems is organized on hierarchical levels (see Fig. 9.1). Typical levels of such a structure are

- central power system control, i.e. load dispatcher
- local power system control, i.e. regional load dispatcher
- station control
- feeder or object control

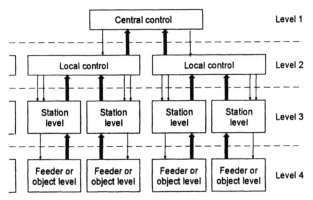

Fig. 9.1:   Hierarchical structure for power system control
➡ data flow,   ➡ command flow

Table 9.1   Typical tasks performed at central and/or local power system control levels

Task	State		
	Norm. op.	Fault	Resto-ration
Exchange of data between computer systems on different power system control levels	•	•[1]	•
Data acquisition and verification	•	•[1]	•
Data concentration	•	•[1]	•
Determination of system topology	•	•[1]	•
Determination of load flow and voltages	•	•[1]	•
Selection of power plants and generator units	•		•
Spinning reserves control	•	(•)	•
High-level frequency/active power regulation	•		
High-level frequency/reactive power regulation	•		
Operational reliability estimation	•		•
Estimation of the effects of power system configuration changes	•		•
Event recording	•	•	•
High-level back-up protection		•	
System-wide confinement of faults		•	
Analysis of further development	•		•
Measurement and recording of selected parameters	•	•	•
Fault analysis			•
Reporting	•		•
Operator interfaces	•	•	•
Self-supervision and diagnosis	•		

[1] Only for protection functions or protection consequences

Data flow mainly from the process via the feeder and station control levels to the local control level and commands in the opposite direction; in most cases, the lowest level has the greatest priority.

9.2.2   Areas of responsibility of the different control levels

The tasks typically performed at the various hierarchical control levels are summarized in Table 9.1. They fall into three time periods and differ considerably within these periods. As a consequence, the requirements to be fulfilled by the communication and processing devices in the three time periods are basically different. The time periods, respectively the corresponding power system states are

- **Normal operation:** The main task consists in the provision at minimum cost (generation and transmission costs) of a dependable supply of electrical energy within the prescribed voltage and frequency limits.

- **Fault:** The task in the event of a fault is to confine its effects which is normally achieved by selectively isolating the afflicted part of the system in the shortest possible time. This task is performed automatically by the protection devices at feeder or object, or possibly at station level. Control of these protection devices at a higher control level will probably become possible in the future.

- **Power system restoration:** The supply must be restored as quickly as possible to consumers isolated by operation of protection devices. Only limited automation of functions of this kind has taken place up to the present and it is to be expected that computer-based systems will be implemented in the foreseeable future.

The requirements to be fulfilled by the communication systems between and within the control levels during the three time periods differ considerably and will presumably grow in future.

Table 9.2 lists typical control, protection and supervision tasks at the feeder and station control levels (according to Fig. 9.1).

9.2.3   Establishment of computer structures in substations

It follows from Table 9.2 that many of the most important functions which have to be carried out quickly at the feeder level only need information available from the feeder itself (e.g. current, voltage, configuration for protection, measurement, metering etc.) and the resulting commands act primarily on the feeder. The actions performed and the measurements are transmitted to the station control level (see Fig. 9.2). This applies especially to the fault state which places high demands on processing speed. The feeder control level obtains amongst other things the settings for the protection devices and details of the system configuration from the station control level. The logical conclusion is therefore to construct switchgear bay units which integrate as far as possible all the functions associated with a bay in one or more microprocessors. Such specifically feeder functions are protection, feeder control and interlocks, measurement, metering and diagnostics. Whether it is possible to implement all these functions in a single processor depends on the power system voltage and operating philosophy, back-up protection design etc. The boundary conditions with regard to the absolute immunity of the protection to interaction and its priority in relation to other functions were already mentioned in Section 9.1 and also apply to the multiple utilization of input variables (e.g. current for protection and measurement).

Table 9.2   Control, protection and supervision tasks at the station control level

Task	State				Level
	Norm. op.	Defective		Resto- ration	
		Dur. fault	After fault		
1	2	3	4	5	6
Exchange of data between computer systems on different power system control levels	•	• [1]	• [1]	•	3 4
Data acquisition, verification and concentration	•	• [1]	• [1]	•	3 4
Generation of alarm signals		•	•		3 4
Feeder protection		•			4
Busbar protection		•			3
Back-up protection		•			3 4
Breaker back-up protection		•			3
High-level supervision of protection settings		•		•	3
Auto-reclosure			•	•	3 4
Automatic switching of supply sources			•	•	3 4
Automatic load-shedding			•	•	3 4

Table 9.2   Continued

1	2	3	4	5	6
Generation control and load dispatching	•		•	•	3
Overload supervision of station plant and lines	•		•	•	3 4
Stability optimization	•		•	•	3
Automation of power system build-up switching procedures				•	3
Determination of switching sequences and interlocks	•			•	3
Event recording	•	•	•	•	3
Detection and recording of disturbances		•	•		3 4
Fault location on overhead lines		•			4
Measurement and recording of analogue signals	•			•	3 4
Voltage regulation	•			•	3
Optimization of transformer loads	•				3
Supervision of auxiliaries	•			•	3
Functional analysis of computer systems	•			•	3
Operator interfaces	•			•	3
Self-supervision and diagnosis	•			•	3 4

[1] Only for protection functions or protection consequences

It also occurs and presumably will be frequently the case during the transition from analogue to digital secondary technology that some switchgear bay functions are performed by a microprocessor and others belonging to the same feeder by conventional units and this necessitates the definition of suitable interfaces. The bus systems used at the feeder control level are of the high-speed parallel kind.

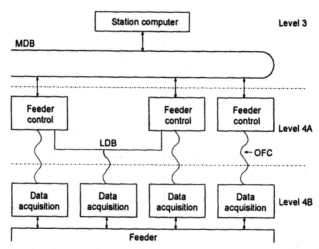

Fig. 9.2: Example of a computer structure for a substation
MDB = main data bus, LDB = local data bus, OFC = optical fibre cables

To what extent the bay units are equipped with their own MMC's (man/machine communication), thus permitting all settings etc., to be performed at the feeder level, or all setting and supervision actions have to be executed on a central MMC at the station control level, is a question of cost and operating philosophy. The tasks at the busbar and station control levels are basically the same and for this reason both busbar protection and station control level process protection (busbar and breaker back-up protection) and interlocking (e.g. bus-tie breaker, isolators) functions. The preceding remarks regarding multiple utilization of input variables and MMC also apply to these units which can also serve as back-up for the bay units.

The communication system between the bay and station control levels is arranged either radially or as a bus. The highest reliability and speed demands it has to fulfil are determined once again by the busbar protection, but synchronization and interlocking functions are similarly critical.

The basic requirements which digital protection and control systems have to fulfil with regard to selectivity, availability, speed etc., are naturally the same as for conventional analogue protection and control devices. Digital techniques, however, offer the additional advantage of extensive self-supervision and therefore open up new avenues for improving maintenance and availability.

Digital measurement and processing systems will come into their own as soon as the input variables are transferred in digital form by the measurement transducers, which will be the case when what are referred to today as unconventional c.t's and v.t's are in general use.

Fully integrated computer structures for substations are gradually being introduced. During the transitory period from analogue relays and conventional secondary technology to a fully digital system, the use of microprocessors will become standard practice in the individual protection, control and metering units which will be applied together with analogue devices.

Independently of its technical design, the computer-based control system receives its input signals from measurement transducers and transmitters and generates corresponding output signals for controlling the plant (Fig. 9.3). The input signals may be analogue, digital or binary states. Analogue variables come mainly from c.t's and v.t's, but may also be signals proportional to physical parameters (e.g. temperature, pressure etc.). Binary states originate primarily from auxiliary contacts on various items of plant, i.e. circuit-breakers, isolators, tap-changers etc. Signals which are already in a digital form are transmitted by digital communications equipment and control keyboards.

The output signals generated by the computer-based protection and control system energize the actuating coils of switches and power transformer tap-changers, change the settings of analogue devices, are displayed on the screens of workstations and registered by recording devices and transmitted to remote locations by communications equipment. In some cases analogue output signals provide for continuous control of plant.

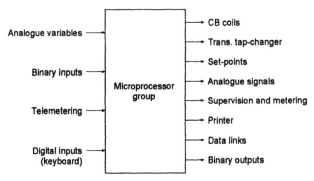

Fig. 9.3:   Input and output signals of a digital protection and control system for a substation

9.3 Special features of digital measurement and control methods

The conversion of an analogue variable to a digital variable is quite different from processing an analogue variable in an analogue device. The major differences are

- Computer-based systems have an easily accessible, programmable and relatively large memory capability which analogue devices do not.
- In the case of fast control systems, the objective is to execute a complete series of operations between individual samples which is then repeated cyclically. This means that computations have to be carried out in a very short time.
- Computer-based systems are able to make better use of analogue and binary state signals by determining their interrelationships.
- Digital signals give no direct information on the behaviour of the input variable between the samples which compared with an analogue measurement necessitates some complicated operations to determine the zero-crossing.

For the above reasons there is little to be gained by simply mimicking analogue protection and/or control devices digitally.

10

A/D Conversion of Input Variables

10.1 Fundamental considerations

A typical procedure for the analogue-to-digital conversion of input variables is give in Fig. 10.1. Continuously varying signals — usually sinusoidal currents and voltages from c.t's and v.t's — are applied to the input of the converter. These, however, may not just consist of the fundamental, but may include superimposed HF interference, harmonics, subharmonics and also a DC component. The general equation for the input signal is thus

$$v(t) = V_1 \cos(\omega_1 t - \alpha_1) + V_0 \, e^{-t/T_a} + \sum_{k=2}^{p} V_k \cos(k\omega_1 t - \alpha_k) \qquad (10.1)$$

where
V_1 = amplitude of the fundamental at rated frequency
$V_2...V_p$ = amplitude of the harmonics k times higher than the rated frequency
V_0 = Initial value of the DC component which decays at a time constant T_a

Of the signal components defined by equation (10.1), only a few are data signals used by the control process. The most important of these is the fundamental with a angular velocity ω_1. In some instances a harmonic may be used (second, third or fifth) and less frequently the damped oscillation of a high frequency component. Before performing an A/D conversion, it is important to ascertain which

Fig. 10.1:   Procedure for converting analogue signals

component in the input variable carries the desired information and therefore should be converted accurately and which is to be classified as interference and should be suppressed.

10.2 Analogue signal filter

Analysis of Fig. 10.2 demonstrates the necessity of filtering analogue signals. It is assumed that the input variable has a constant amplitude V_ϕ upon which a sinusoidal signal v_p with the angular velocity ω_1 is superimposed. If the frequency of this signal equals the sampling rate of the A/D converter (Fig. 10.2a), the converter would produce a series of samples with the constant value $V_\phi + V_p$, i.e. there would be a constant error equal to V_p. The error assumes other values in the range $\pm V_p$ for sampling instances at other points on the time function V_p. For a sampling rate which is twice the frequency of the component V_p (Fig. 10.2b) and thus also of the discrimination frequency determined by Shannon for curves

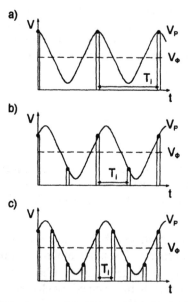

Fig. 10.2:   Influence of the sampling rate on the discrimination error

varying as a function of time, the errors due to the DC component can be eliminated, for example, by averaging a given number of samples. The sampling rate is too low, however, should both the constant term V_ϕ and the amplitude of the function v_p be reproduced.

By choosing a sampling rate four times the component V_p (Fig. 10.2c), the samples obtained evaluate both the initial value V_0 and the amplitude V_p. However, as can be seen from Fig. 10.2a, an irreversible error is introduced, should the input variable contain an oscillation at a higher frequency than the sampling rate. To avoid difficulties of this kind, the following have to be chosen carefully

- sampling rate of the A/D converter where $f_i = 1/T_i$
- characteristic of the bandpass filter for suppressing frequencies higher than $0.5 f_i$

Examination of the influence of both parameters on the signal discrimination error discloses the following

- The sampling rate f_i must be at least four times higher than the frequency of those components of the input variable which have to be accurately reproduced for further processing, i.e. $f_i \gg 4f_p$.
- The characteristic of the bandpass filter must be such that all components of the input variable of frequency f_p are accepted and all others with frequencies higher than $(f_i - f_p)$ are rejected. In practice this means that the rejection frequency $f_c = \left(f_i - f_p\right)/3$.

A filter of this kind is called an anti-aliasing filter.

Should it prove impossible to fulfil these conditions or extremely high accuracy be desired, a detailed analysis must be conducted which takes account of the quantitative aspects of signal discrimination.

The analogue input variables applied to the A/D converter must be adjusted to lie within its operating range. For example, an A/D converter with a voltage output signal usually has a maximum signal value of 10 V and one with a current output 5 mA.

The above bandpass filter is referred to a an "anti-aliasing" filter.

10.3 A/D signal conversion

Mainly single A/D converters with a multiplexer for several analogue input variables are used in practical applications. The purpose of the multiplexer is to apply the input variables to the converter in turn for sampling.

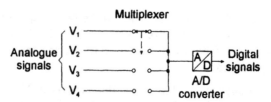

Fig. 10.3:   Principle of an A/D converter in combination with a multiplexer

Multiplexers can be divided into two groups. The first group includes simple solid-state switches which sample the input variables by connecting them in sequence to the converter. The principle is shown symbolically in Fig. 10.3. The individual signals are scanned at an interval equal to the minimum processing time of the A/D converter. The signal evaluation circuits must take this into account.

The second group of multiplexers stores the instantaneous values of all the input variables at the beginning of every cycle. This is simulated in Fig. 10.4 by briefly closing the contacts K_1. The values stored are then applied in sequence to the A/D converter by the contact K_2. As before, conversion takes time, but in this case the values processed all relate to the same instant in time and it is therefore permissible to compare them. The cycle starts again after all the samples have been scanned. Although represented symbolically by contacts in Figures 10.3 and 10.4, all the operations are performed in practice by solid-state devices.

There are many ways of performing the A/D conversion function, but in all cases two basic parameters are important

- the sampling rate $f_i = 1/T_i$ which is directly related to the optimization of the analogue filter
- the word length (i.e. the number of Bits) used to define each sample of the analogue signal

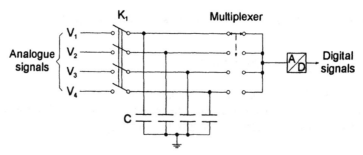

Fig. 10.4:   Coordination of an A/D converter and a multiplexer with memory

Given a word length of m Bits (excluding the sign Bit), the maximum digital range is

$$N = 2^m - 1 \qquad (10.2)$$

i.e. the difference between two adjacent digital values of the signal is

$$\Delta B_e = \frac{B_e}{N} \qquad (10.3)$$

where
B_e = maximum signal range

If at the instant of sampling the value of the signal lies between two digital values, the sample has to be rounded. The alternatives are to round upwards or downwards, for which the maximum errors become $\pm 0.5\ \Delta B_e$ and ΔB_e respectively.

The range of B_e must not be less than the maximum value to be expected for the signal concerned, while the maximum error has to be referred to the minimum value of the signal to be measured and its specified accuracy. The following conditions result

$$B_e \geq X_{max}$$
$$\varepsilon \geq \frac{\Delta B_e}{2\,X_{min}} \qquad (10.4)$$

where
X_{max}, X_{min} = maximum and minimum values of the signal
ε = permissible relative error of the lowest signal value

Taking equation (10.3) into account

$$N \geq \frac{X_{max}}{2\,\varepsilon\,X_{min}} \qquad (10.5)$$

which enables the word length, i.e. the number of Bits m, to be calculated.

The following examples for the measurement of current and voltage can be used in order to check whether the protection requirements can be fulfilled.

Example 10.1: Measurement of voltage

The range necessary for measuring the r.m.s. values of secondary voltages which have to be known accurately is $0.1 \cdot 1000$ V to $2 \cdot 1000$ V. The amplitude error at the minimum value of the input variable should not exceed 5 %.

Thus

$$X_{max} = 2 \cdot 100\sqrt{2} \text{ V}; \quad X_{min} = 0.1 \cdot 100\sqrt{2} \text{ V}; \quad \varepsilon = 0.05$$

and therefore according to equation (10.5), $N \geq 200$ and according to equation (10.2), $m = 8$.

Example 10.2: Measurement of current

The r.m.s. value of a c.t. secondary current is 1 A and this should be measured to an accuracy of 1 %. The r.m.s. secondary fault current is 40 A and it is to be expected that it will also contain a DC component which will increase this value by a factor of 2. Thus

$$X_{max} = 40 \cdot 2\sqrt{2} \text{ A}; \quad X_{min} = \sqrt{2} \text{ A}; \quad \varepsilon = 0.01$$

and therefore $N \geq 4000$ and $m = 12$.

A wide measuring range is therefore essential, especially for current signals, which results in long Bit words. This difficulty can be overcome by applying what is referred to as dynamic measuring range adjustment which on a normal measuring instrument would be equivalent to automatically changing the scale. Whatever the method, the output of an A/D converter is digital signals in the form of words of corresponding length and the point at which digital processing starts.

10.4 Verification

Verification (Fig. 10.1) is the operation of checking whether the digital form of the input variable contains an error. If an error is found, then the verification function should correct it if possible. Error correction must be based on the redundancy of data in relation to the minimum amount of essential information. The excess information (redundancy) can be determined in one of the following ways

- repeated measurement of the input variable or measurement of additional variables which permit the variable in question to be determined indirectly
- utilization of information available on the range within which the variable could be or its relationship to other variables
- by expanding the information area and forming "forbidden zones"

The most common causes of errors are

- obvious errors due to a fault in the secondary circuits of c.t's and v.t's (e.g. loss of voltage signal because of a blown fuse) or failure of a communications channel

- errors in signal transducers and transmitters due to overload, e.g. the saturation of c.t's
- Bit errors in the communication of digital signals

The methods of state estimation which have been common practice for years in power system control and include the detection and correction of erroneous data are becoming established at substation level. The latter now forms the lowest level of power system state estimation. The following examples of error correction procedures are based on Fig. 10.5.

Figure 10.5a illustrates a method of verifying the current measurements in three feeders with the aid of Kirchoff's law, i.e. the sum of the three currents has to equal zero.

$$\underset{\cdot}{I_1} + \underset{\cdot}{I_2} + \underset{\cdot}{I_3} \overset{?}{=} 0$$

This, of course, must also apply under normal load conditions.

The second method in Fig. 10.5b makes use of an excess of signals. The secondary voltages $\underline{U}_{L1}$, $\underline{U}_{L2}$, $\underline{U}_{L3}$ and $\underline{U}_A$ are measured and assuming a standard v.t.

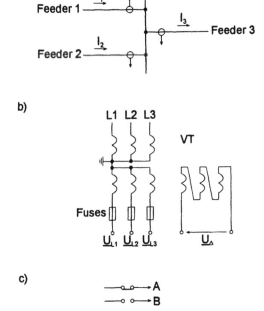

Fig. 10.5: Simple methods of signal verification

ratio, the following condition must be satisfied

$$\underline{U}_{L1} + \underline{U}_{L2} + \underline{U}_{L3} = \sqrt{3}\ \underline{U}_{\Delta}$$

An error exists whenever this condition is not fulfilled.

The third very simple example verifies the state of a circuit-breaker by comparing the signals from two auxiliary contacts (Fig. 10.5c). Both contacts in the same position indicates an error, thus it is monitored that $A \neq B$. For purposes of analysis, A respectively B = logical '1' when it is closed and logical '0' when it is open.

The formation of "forbidden zones" as a method of verification involves the establishment of codes when estimating the transmission errors of the individual signal Bits. The simplest code for this purpose is the use of a parity Bit, i.e. a Bit is added to every word with a value to make the total number of Bits with a value of 1 even. The parity Bits for the numbers 0 to 9, for example, are

Number	Word	Parity Bit
0	0000	0
1	0001	1
2	0010	1
3	0011	0
4	0100	1
5	0101	0
6	0110	0
7	0111	1
8	1000	1
9	1001	0

This code can detect single word errors but is powerless to detect double errors.

A much more effective but complicated code is called the "same index code". It relies on the fact that the same number of ones must exist in every word. Assuming a word has m Bits and the constant number of ones equals k, the number of digital states which can be expressed in this way is

$$\binom{n}{k} = \frac{n!}{k!}(n-k)!$$ (10.6)

By comparison, the number of digital states which can be expressed by an non-coded word with m Bits is $2^m - 1$.

The "two-out-of-five" code is also used frequently in practice, i.e. the number of ones in a five Bit word is fixed at two. The following results for the first ten numbers

Number	Word
0	10000
1	00011
2	00101
3	00110
4	01001
5	01010
6	01100
7	10001
8	10010
9	10100

This code detects the majority of multiple errors as well as single errors. Note, however, that the range of the "forbidden zone" is twice as large as the "permissible zone". This is why the possibilities of writing many state numbers are reduced.

11

Digital Signal Conditioning

11.1 General remarks on digital filtering

Signal conditioning is the function of preparing a signal for processing by the application algorithms. In the case of the control processes in power supply substations, this involves primarily digital filtering and less frequently correlation or the determination of the symmetrical components.

Filters permit desired parts of the signal in a particular frequency band to pass and prevent from passing, or attenuate as far as possible, all other parts of the signal. As with analogue filters, digital filters can be divided into the following groups

- low-pass
- bandpass
- high-pass
- band rejection

The frequency response of the above types of filters is shown in Fig. 11.1 and that of a low-pass filter more accurately in Fig. 11.2. Ideally, filters should have a flat response in the admittance band and steep edges at the cut-off frequencies. Another important parameter with respect to digital filters in communications is their phase-shift/frequency response which should be as linear as possible. While this requirement is less important in protection applications, a filter's behaviour in

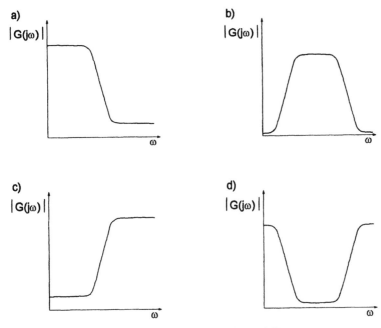

Fig. 11.1:   Frequency response of the basic kinds of filters
              a) low-pass,   b) bandpass,   c) high-pass,   d) band rejection

relation to time is. The impulse response of a filter is also important and should decay quickly without overshoot.

There are three basic criteria for appraising the performance of digital filters for protection and control applications:

1. transmission of the desired components of the signal with adequate accuracy and efficient elimination of all noise

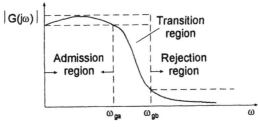

Fig. 11.2:   Frequency response of a low-pass filter

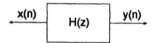

Fig. 11.3:   Symbol for a digital filter

2. fast stabilization of the output signal following a dynamic change of the input signal
3. minimum use of processing capacity, i.e. a minimum of computations

Unfortunately, these requirements oppose each other to some extent and therefore a filter is always a compromise.

A characteristic of all analogue filters is that their impulse response is infinitely long even though it does approach zero with time. There are, however, two kinds of digital filters, those with

- infinite impulse response
- finite impulse response

A typical algorithm of an **infinite impulse response ("recursive") filter** is of the form

$$y(n) = \sum_{k=0}^{p} a_k\, x(n-k) - \sum_{k=1}^{r} b_k\, y(n-k) \qquad (11.1)$$

where

$y(n)$ = n th. output signal sample
$x(n)$ = n th. input signal sample
a_k, b_k = constants

A filter of this kind, the symbol of which is shown in Fig. 11.3, generates the individual samples of the output signal as a weighted sum $(p + 1)$ of the preceding samples of the input and output signals.

A **finite impulse response ("non-recursive") filter** generates its output signals samples solely from the input signal samples. Its algorithm thus has the form

$$y(n) = \sum_{k=0}^{p} a_k\, x(n-k) \qquad (11.2)$$

The term "data window" has been created for filters of this kind which is determined by the following relationship

$$T_w = (p+1)T_i \qquad (11.3)$$

The synthesis of both kinds of filters is dealt with in detail in the following Sections.

11.2 Synthesis of infinite impulse response filters

A digital filter can be described by its frequency transfer function, the determination of which is based on the derivation of the z transformations on both sides of the expression (11.1) [C.17]. This is simple in consideration of the fact that a signal delay of one sample is equivalent to multiplying it by z^{-1}. If therefore $Y(z)$ is the transformation of $y(n)$, then $z^{-1}Y(z)$ is the transformation of $y(n - 1)$. Having determined the z transformation of equation (11.1), the transfer function becomes

$$\frac{Y(z)}{X(z)} = H(z) = \frac{\sum_{k=0}^{p} a_k z^{-1}}{1 + \sum_{k=1}^{r} b_k z^{-k}} \qquad (11.4)$$

The frequency response of the filter is obtained by substituting $\exp(-j\omega T_i)$ for z^{-1}. For an infinite impulse response filter according to equation (11.4), the frequency response is given by

$$H^*(j\omega) = \frac{\sum_{k=0}^{p} a_k \exp(-jk\omega T_i)}{1 + \sum_{k=1}^{r} b_k \exp(-jk\omega T_i)} \qquad (11.5)$$

On the basis of the algorithm according to equation (11.1), calculating and plotting the frequency response of the filter does not present a problem. The synthesis of the filter, however, is an inverse operation, i.e. the coefficients a_k and b_k have to be determined and equation (11.1) generated on the basis of a given characteristic.

There are many ways to synthesize an infinite impulse response filter (see Appendix C). One of the simpler ones based on the transformation of the transfer function of an analogue low-pass filter into a digital filter is presented below. This method has two advantages, firstly the theory of analogue low-pass filters has been exhaustively treated and standard approximations such as those by Butterworth, Tschebyscheff and Bessel are readily available. There are also many standard works containing their transfer functions and frequency responses. Secondly, the method is extremely simple to apply, although the results obtained are not always ideal.

11.2.1 Synthesis of a digital low-pass filter

Assuming an analogue low-pass filter with a transfer function G(s) and a cut-off angular velocity ω_{ga}, the problem is to obtain a digital filter with a similar characteristic, but with a cut-off frequency ω_{gd}.

The procedure consists in substituting the corresponding z operator for the operator s in the expression G(s), i.e.

$$s \Rightarrow \frac{A(1-z^{-1})}{1+z^{-1}} \tag{11.6}$$

The factor A is determined by the condition that the values of the cut-off angular velocity in the transfer functions for analogue and digital filters must be the same. Thus from equation (11.6)

$$\omega_{ga} = \frac{A\left|1-\exp\left(-j\omega_{gd}\,T_i\right)\right|}{\left|1+\exp\left(-j\omega_{gd}\,T_i\right)\right|} \tag{11.7}$$

and therefore

$$A = \omega_{ga}\,\cot\left(\frac{\omega_{gd}\,T_i}{2}\right) \tag{11.8}$$

Example 11.1

The problem is to synthesize a digital low-pass filter having a cut-off angular velocity of $\omega_{gd} = 628$ from a second order Butterworth low-pass filter with a cut-off angular velocity of $\omega_{gd} = 1$ and the transfer function

$$G(s) = \frac{1}{s^2 + \sqrt{s} + 1}$$

The sampling rate $f_i = 1/T_i$ equals 600. The factor A according to equation (11.8) becomes

$$A = \cot 0.523 = 1.733$$

and the transfer function of the digital filter

$$H(z) = \left\{ \left[\frac{1.733(1-z^{-1})}{(1+z^{-1})} \right]^2 + \sqrt{2}\left[\frac{1.733(1-z^{-1})}{(1+z^{-1})} \right] + 1 \right\}^{-1}$$

and also

$$\frac{Y(z)}{X(z)} = H(z) = \frac{1 + 2z^{-1} + z^{-2}}{155z^{-2} - 4z^{-1} + 8.45}$$

Considering that the multiplication by z^{-1} is equivalent to delaying the signal by a sample, it is permissible to write the following

$$8.45y(n) - 4y(n-1) + 155y(n-2) = x(n) + 2x(n-1) - x(n-2)$$

and also

$$y(n) = 0.188x(n) + 0.237x(n-1) + 0.118x(n-2) +$$

$$+ 0.473y(n-1) - 0.183y(n-2)$$

It is thus possible to derive a complete algorithm of an infinite impulse response filter with the desired characteristic and cut-off frequency. It is also necessary, however, to determine the behaviour of the filter in relation to time by checking its impulse and unit step responses which can be accomplished by applying simple simulation procedures.

11.2.2 Synthesis of a digital high-pass filter

Assuming an analogue low-pass filter with a transfer function $G(s)$ and a cut-off angular velocity ω_{ga}, the problem is to derive the algorithm for a digital high-pass filter with a cut-off frequency ω_{gd}. This time the following function for the z operator is substituted for the s operator in the transfer function $G(s)$

$$s \Rightarrow \frac{B(1 + z^{-1})}{(1 - z^{-1})} \tag{11.9}$$

Following a similar approach as for the synthesis of the low-pass filter, it can be show that the factor B has the following form

$$B = \omega_{ga} \tan\left(\frac{\omega_{gd} T_i}{2}\right) \tag{11.10}$$

All other operations are similar to those for the preceding filter.

11.2.3 Synthesis of a digital bandpass filter

Once again assuming an analogue low-pass filter with a transfer function $G(s)$ and a cut-off angular velocity ω_{ga}, the problem is to derive the algorithm for a digital bandpass filter with the cut-off frequencies ω_{d1} and ω_{d2}.

The problem is solved in two steps. In the first step, the analogue low-pass filter is transformed to an analogue bandpass filter by carrying out the following substitution

$$s \Rightarrow c_1 \left(s_1 + \frac{\omega_0^2}{s_1} \right) \tag{11.11}$$

where

s_1 = new form of the Laplace operator

ω_0 = constant representing the angular velocity of the characteristic displacement

In the second step, the analogue bandpass filter is transformed to a digital bandpass filter by making the substitution

$$s_1 \Rightarrow \frac{C_2 \left(1 - z^{-1}\right)}{\left(1 + z^{-1}\right)} \tag{11.12}$$

Both steps can be performed simultaneously by substituting

$$s \Rightarrow \frac{C \left(1 - 2\alpha z^{-1} + z^{-2}\right)}{\left(1 - z^{-2}\right)} \tag{11.13}$$

where

$$C = \omega_{ga} \cot \left[\frac{T_i \left(\omega_{d2} - \omega_{d1}\right)}{2} \right] \tag{11.14}$$

$$\alpha = \cos \left[\frac{T_i \left(\omega_{d1} + \omega_{d2}\right)}{2} \right] \Bigg/ \cos \left[\frac{T_i \left(\omega_{d2} - \omega_{d1}\right)}{2} \right] \tag{11.15}$$

The remaining operations are identical to the previous cases with the exception that since the filter is a second order filter more time is required for the calculations.

11.2.4 Synthesis of a digital band rejection filter

The substitution which has to be made to obtain the algorithm for a digital band rejection filter with the cut-off frequencies ω_{d1} and ω_{d2} from a given analogue low-pass filter with a transfer function $G(s)$ and a cut-off angular velocity ω_{ga} is as follows

$$s \Rightarrow \frac{D\left(1 - z^{-2}\right)}{\left(1 - 2\alpha z^{-1} + z^{-2}\right)} \qquad (11.16)$$

where

$$D = \omega_{ga}\ \tan\left[\frac{T_i\left(\omega_{d2} - \omega_{d1}\right)}{2}\right] \qquad (11.17)$$

In all cases, the digital transfer function H(z) is obtained by substituting a corresponding function for the s operator in the transfer function G(s). The algorithm for the filter can then be simply derived from the transfer function H(z).

While having many advantages, the synthesis of digital filters described above unfortunately also have some disadvantages. One is that the characteristics of the algorithms are not an optimum, mainly with regard to their dynamic behaviour. Another is that the samples x and y have to be individually multiplied by the coefficients a_k and b_k which takes processing time.

11.3 Synthesis of finite impulse response filters

Filters with a finite impulse response are described by the following algorithm

$$y(n) = \sum_{k=0}^{p} a_k\ x(n-k) \qquad (11.18)$$

The procedure for determining the transfer function for this filter is similar the previous one in that the z transformations of both sides of the expression (11.18) are derived first and then the delay by one sample taken into account by multiplying by z-1. The following transfer function for the filter then results

$$\frac{Y(z)}{X(z)} = H(z) = \sum_{k=0}^{p} a_k\ z^{-k} \qquad (11.19)$$

The frequency response of the filter is obtained by substituting the operator $\exp(-jT_i)$ for z^{-1}

$$H^*(j\omega) = \sum_{k=0}^{p} a_k\ \exp\left(-jk\omega T_i\right) \qquad (11.20)$$

The phase and amplitude characteristics can be determined from this.

Analysis of equation (11.18) discloses that the frequency response of the filter can be obtained much more simply, because the algorithm (11.18) is nothing other than the integration according to Euler of the signal $x(\tau)$ multiplied by the

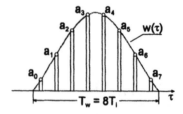

Fig. 11.4:   Distribution of factors in the data window

factors a_k which are distributed as shown in Fig. 11.4 across the data window. The algorithm (11.18) is thus the digital form of the integral

$$y(n) = \sum_{k=0}^{p} a_k \, x(n-k) \Leftrightarrow y(t) = \frac{1}{T_i} \int_{-\infty}^{\infty} x(\tau) \, w(t - \tau) d\tau \qquad (11.21)$$

The corresponding graphical response is given in Fig. 11.5.

The above integral is the convolution of the input signal $x(\tau)$ and the function $w(\tau)$ which form the envelope of the data window. Algorithm (11.18) is by contrast the digital version of this integral.

Taking account of the fact that the Fourier transformation of the convolution equals the product of the transformations, then

$$Y(j\omega) = X(j\omega) \, W(j\omega) \qquad (11.22)$$

where

$Y(j\omega)$, $X(j\omega)$ = Fourier transformations, i.e. the spectrum of the signals $y(\tau)$, $x(\tau)$

$W(j\omega)$ = Fourier transformation of the data window $w(\tau)$ divided by the impulse duration T_i

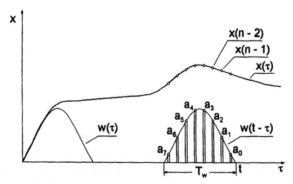

Fig. 11.5:   Graphical representation of algorithm (11.18)

The following two conclusions can be drawn from this

1. The data window $w(\tau)$ is the impulse response (weighting function) of the filter divided by the impulse duration.
2. The Fourier transformation of the data window $w(\tau)$ is the filter spectrum divided by T_i.

These conclusions have practical consequences, since instead of laborious calculations on the basis of equation (11.20), the spectrum of the filter can be determined much more simply via the Fourier transformation of the data window $w(\tau)$. It should be noted, however, that this simplified method does involve a certain inaccuracy, because it does not take the digital procedure for determining the convolution integral into account. As a consequence, the side-bands of the spectrum representing the results which are displaced by $w_i = 2\pi/T_i$ are ignored. The influence of the side-bands, however, is only slight for short impulse durations T_i, i.e. correspondingly high impulse frequencies ω_i.

With regard to their centers, the data windows $w(\tau)$ are frequently symmetrical or alternating (even or odd) functions and it is therefore an advantage when determining the Fourier transformations to shift the functions so that their centers coincide with the origin of the coordinates as shown in Fig. 11.6. A shifted function is described by $w_r(\tau) = W\!\left(\tau + T_w/2\right)$ from which

$$W(j\omega) = W_r(j\omega)\exp\!\left(\frac{-j\omega T_w}{2}\right) \tag{11.23}$$

The synthesis of filters with a finite impulse response is based on the determination of the factors a_k which describe the amplitude characteristic. There are many more or less complicated methods of synthesizing filters [C.1, C.17]. A "demonstration synthesis" is presented below which is based on the analysis of the characteristic properties of filters with frequently used data windows.

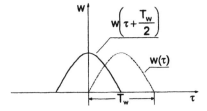

Fig. 11.6:   Shifting a data window

11.3.1 Low-pass filter with finite impulse response

The most commonly used low-pass filter has a rectangular data window, i.e. all the factors a_k have the value 1 (Fig. 11.7). The Fourier transformation of this window which is equivalent to the amplitude spectrum is of the form

$$W_r(j\omega) = \frac{\dfrac{T_w}{T_i}\sin\left(\dfrac{\omega T_w}{2}\right)}{\dfrac{\omega T_w}{2}} \tag{11.24}$$

and therefore

$$W(j\omega) = \frac{(p+1)\left[\sin\left(\dfrac{\omega T_w}{2}\right)\exp\left(\dfrac{-j\omega T_w}{2}\right)\right]}{\dfrac{\omega T_w}{2}} \tag{11.25}$$

The amplitude spectrum given by equation (11.25) is shown graphically in Fig. 11.8. The characteristic properties of a low-pass filter are obvious, although they are adversely influenced by the side-bands at frequencies higher than $2\pi/T_w$. An important advantage of this filter is that it is simple to calculate, because the digital algorithm is of the form

$$y(n) = \sum_{k=0}^{p} x(n-k) \tag{11.26}$$

which only requires p additions. If the data window contains a high number of samples, the further reduction of the number of operations can be achieved by applying the infinite impulse response relationship for algorithm (11.26). By inserting y(n - 1) for the algorithm, the following regularity becomes evident

$$y(n) = y(n-1) + x(n) - x(n-p-1) \tag{11.27}$$

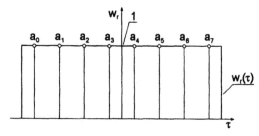

Fig. 11.7: Rectangular data window

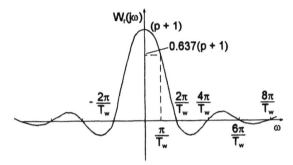

Fig. 11.8:   Amplitude spectrum of a filter with rectangular data window

Thus the calculation of every subsequent sample only requires two addition or subtraction operations irrespective of the number of samples per data window.

Filters with rectangular data windows are used most, because they have excellent filtering properties and yet are simple to calculate. Where it is considered necessary to reduce the influence of the side-bands of the frequency response, it must be born in mind that this is mainly due to the step-change of the function $w(\tau)$ at the edges of the window. Attenuating this influence reduces the side-bands, but increases the width of the admittance spectrum. A typical example is a low-pass filter with a triangular data window (Fig. 11.9) the amplitude spectrum of which is given by the following relationship

$$W_r(j\omega) = \frac{(p+1)\left[1-\cos\left(\dfrac{\omega T_w}{2}\right)\right]}{\left(\dfrac{\omega T_w}{2}\right)^2} \tag{11.28}$$

The corresponding response curve in Fig. 11.10 shows clearly that the admittance spectrum is twice as wide as for a rectangular data window and the side-bands are negligibly small.

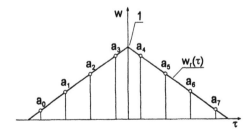

Fig. 11.9:   Triangular data window

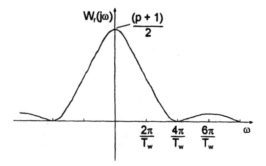

Fig. 11.10: Amplitude spectrum of a filter with triangular data window

The data windows of low-pass filters can have different shapes, e.g. the Blackman, Hamming or cosine data window. All of these windows have in common that the phase-shift caused by the filter varies linearly with frequency, i.e. an input signal with the angular velocity ω_x appears at the output of the filter with a corresponding amplitude and a phase-shift α_x where

$$\alpha_x = -\frac{\omega_x\, T_w}{2} \tag{11.29}$$

Example 11.2

A discrete signal defined by the relationship

$$x(n) = X_0 + X_1 \cos\omega_1 nT_i + X_2 \cos 2\omega_1 nT_i$$

is applied to a digital low-pass filter having a rectangular data window and the factors $a_k = 1$. There are 6 samples in the data window of width $T_w = \omega_1/2$, i.e. $T_w/T_i = (p + 1) = 6$. The response of the steady-state filter output signal has to be determined.

The amplitudes are

$$W_r(j0) = p+1 = 12$$

$$W_r(j\omega_1) = 0.637(p+1) = 7.63$$

$$W_r(j2\omega_1) = 0$$

The phase-shift for $\omega = \omega_1$ is

$$\alpha_1 = -\frac{\omega_1\, T_w}{2} = \frac{\pi}{2}$$

Thus the steady-state output response is

$$y(n) = 12X_0 + 7.63X_1 \cos\left(\omega_1 n T_i - \frac{\pi}{2}\right)$$

11.3.2 Bandpass filter with finite impulse response

a) Sine/cosine data windows

Bandpass filters with data windows which are part of sine or cosine functions are especially important, because they are useful for filtering specific frequencies out of signals. The sections of periodic functions are chosen such that the sine function reaches its minimum and the cosine function its maximum in the center of the data window. This is the case in Fig. 11.11 which shows the two most common situations of the data window corresponding to one period of the fundamental frequency of the sine, respectively cosine function.

Independently of the width of the window, the amplitude spectrums for sine and cosine data windows can be relatively simply determined from the following equations

- Sine data window

$$W_{rs}(j\omega) = j\frac{p+1}{2}\left\{\frac{\sin\left[(\omega - \omega_0)\frac{T_w}{2}\right]}{(\omega - \omega_0)\frac{T_w}{2}} - \frac{\sin\left[(\omega + \omega_0)\frac{T_w}{2}\right]}{(\omega + \omega_0)\frac{T_w}{2}}\right\} \tag{11.30}$$

- Cosine data window

$$W_{rc}(j\omega) = \frac{p+1}{2}\left\{\frac{\sin\left[(\omega - \omega_0)\frac{T_w}{2}\right]}{(\omega - \omega_0)\frac{T_w}{2}} + \frac{\sin\left[(\omega + \omega_0)\frac{T_w}{2}\right]}{(\omega + \omega_0)\frac{T_w}{2}}\right\} \tag{11.31}$$

where
$W_{rs}(j\omega)$ = amplitude spectrum of the sine data window
$W_{rc}(j\omega)$ = amplitude spectrum of the cosine data window
$\omega_0 = \dfrac{2\pi}{T_0}$ = angular velocity of sine and cosine functions

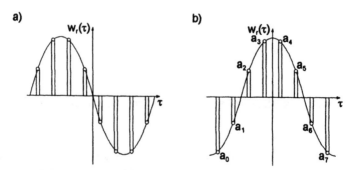

Fig. 11.11: a) sine data window and b) cosine data window

For the cases shown in Fig. 11.11, $\omega_0 = 2\pi/T_w$; the amplitude spectrums determined for them using equations (11.30) and (11.31) can be seen in Fig. 11.12. Figure 11.13 gives examples of typical amplitude spectrums for different relationships between window width and the periods of sine and cosine functions, the data windows being on the left and the respective amplitude spectrum on the right.

Filters with the characteristics shown in Figures 11.12 and 11.13 accept frequencies with angular velocities of ω_0.

Comparison of the filter properties of sine and cosine data windows indicates that for window widths $T_w < T_0$, the sine data window exhibits good attenuating properties for frequency components with angular velocities $\omega < \omega_0$ but poor performance at frequencies higher than ω_0. The circumstances are reversed for window widths $T_w \geq T_0$.

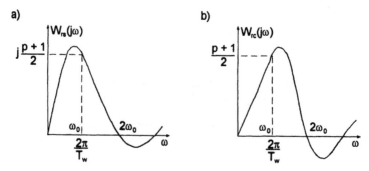

Fig. 11.12: Frequency responses of the functions in Fig. 11.11
            a) sine data window,   b) cosine data window

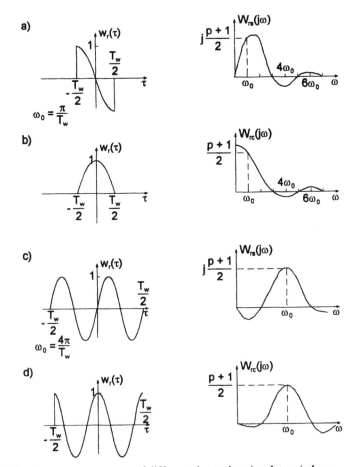

Fig. 11.13: Frequency responses of different sine and cosine data windows

An important characteristic of both sine and cosine data windows is that the phase-shift caused by the filters is $\pi/2$ for components of all frequencies. the phase-shift is give by the following equations

$$\alpha_s(j\omega) = -\frac{\omega T_w}{2} + \frac{\pi}{2} \qquad (11.32)$$

$$\alpha_c(j\omega) = -\frac{\omega T_w}{2} \qquad (11.33)$$

where $\alpha_s(j\omega)$ and $\alpha_c(j\omega)$ are the phase-shifts caused by the sine and/or cosine windows for the signal components with the angular velocity ω.

Digital filters with the data windows described above can be designed with the aid of the following algorithm

$$y(n) = \sum_{k=0}^{p} a_k \, x(n-k) \tag{11.34}$$

where for the sine data window

$$a_k = \sin\left[\omega_0 T_i\left(\frac{p}{2}-k\right)\right] \tag{11.35}$$

and for the cosine data window

$$a_k = \cos\left[\omega_0 T_i\left(\frac{p}{2}-k\right)\right] \tag{11.36}$$

The execution of the algorithm of equation (11.34) involves $(p + 1)$ multiplications which take some considerable time. By exploiting the properties of the sine and cosine functions, it is possible to reduce the number of multiplications by a factor of 2. This is achieved by grouping the terms in algorithm (11.34) such as to take advantage of those values of the factors a_k which are repeated and is based on the fact that the following can be written for a sine data window

$$y_s(n) = \sum_{k=0}^{\frac{(p-1)}{2}} \left[x\left(n-\frac{p}{2}+k+0.5\right) - x\left(n-\frac{p}{2}-k-0.5\right)\right] \times$$
$$\times \sin\omega_0 T_i(k+0.5) \tag{11.37}$$

respectively the cosine window

$$y_c(n) = \sum_{k=0}^{\frac{(p-1)}{2}} \left[x\left(n-\frac{p}{2}+k+0.5\right) + x\left(n-\frac{p}{2}-k-0.5\right)\right] \times$$
$$\times \cos\omega_0 T_i(k+0.5) \tag{11.38}$$

Further simplifications are possible where the data window is wider than half a period of the frequency ω_0. In extreme case, grouping the terms can result in the number of absolutely essential multiplications being reduced to the sampling rate per quarter period of the frequency ω_0 which is $(p + 1)T_0/4T_w$, where $T_0 = 2\pi/\omega_0$. Providing the number of samples per data window is relatively low, these reductions of computation requirement are fully adequate, but if $(p + 1)$ samples is large and especially for filters with sine and/or cosine data windows, it is better to express algorithm (11.34) in the infinite impulse response form. This necessitates using the complex form of the sine functions, i.e.

$$g(n) = \sum_{k=0}^{p} x(n-k) \exp\left[j\left(\frac{p}{2}-k\right)\omega_0 T_i\right]$$ (11.39)

Filters with sine and cosine windows can now be written

$$y_s(n) = \text{Im}(g(n)) \qquad\qquad y_c(n) = \text{Re}(g(n))$$

There are many methods of determining the form of infinite impulse response algorithms, only one of which is presented below for the case that the length of the data window T_w is a multiple of the half-period $T_0/2$ plus one sample, i.e.

$$T_w = \frac{kT_0}{2} + T_i = (p+1)T_i$$

From this expression, which enables $g(n-1)$ to be calculated, and equation (11.39)

$$g(n) = \left[g(n-1) - x(n-p-1)\exp\left(\frac{-j\omega_0 T_p}{2}\right)\right]\exp(-j\omega_0 T_i) +$$
$$+ x(n)\exp\left(\frac{-j\omega_0 T_i p}{2}\right)$$ (11.40)

The advantage of this form is that $\exp(j\omega_0 T_i p/2)$ can have the values ±1 or $\pm j$ which means that the calculation of apparent and real components of $g(n-1)$ requires the multiplication of four real numbers.

Example 11.3

The window width T_w in digital protection devices is frequently required to equal almost half the angular velocity ω_0. Assuming $T_w = (p+1)T_i = \frac{T_0}{2} + T_i$,

$$\exp\left(\frac{j\omega_0 T_i p}{2}\right) = j$$

and therefore

$$g(n) = \left[g(n-1) + jx(n-p-1)\right]\exp(-j\omega_0 T_i) + jx(n)$$

From this follows that

$$y_s(n) = \text{Im}\{g(n)\} = \left[y_s(n-1) + x(n-p-1)\right]\cos\omega_0 T_i -$$
$$- y_c(n-1)\sin\omega_0 T_i + x(n)$$

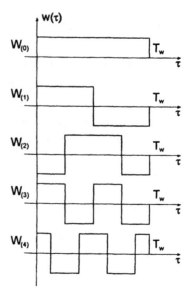

Fig. 11.14: The first five Walsh functions

and

$$y_c(n) = \text{Re}\{g(n)\} = [y_s(n-1) + x(n-p-1)]\sin\omega_0 T_i +$$
$$+ y_c(n-1)\cos\omega_0 T_i$$

b) Walsh function data windows

In automation, filter data windows in the form of Walsh functions are frequently used. The first four of these are shown in Fig. 11.14. Their advantage is principally that they can only assume the values +1 and −1 and thus significantly reduce the computation requirement.

The previous Section described the data window as a zero order Walsh function, since it produces a typical low-pass filter (see Figures 11.7 and 11.8). The data window in Fig. 11.15, on the other hand, corresponds to a Walsh function of the first order with its center shifted to the origin of the coordinates. The spectrum of the data window is described by the following relationship

$$W_r(jw) = j\frac{T_w}{T_i}\left[\frac{1-\cos\left(\dfrac{\omega T_w}{2}\right)}{\dfrac{\omega T_w}{2}}\right] \qquad (11.41)$$

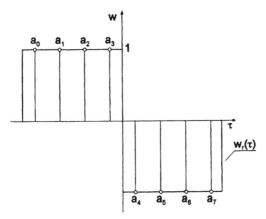

Fig. 11.15: A data window corresponding to a Walsh function of the first order

The curve of the above equation is given in Fig. 11.16.

The digital execution of the filter with the above data window is relatively simple, since

$$y(n) = \sum_{k=0}^{\frac{p-1}{2}} x(n-k) - \sum_{k=\frac{p+1}{2}}^{p} x(n-k) \tag{11.42}$$

and therefore no multiplication operations are required, only p addition/subtraction operations. Where the number of samples per data window is very high, a certain reduction of the computation requirement can be achieved

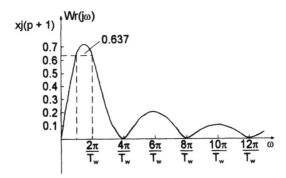

Fig. 11.16: Amplitude spectrum of the data window in Fig. 11.15

using the infinite derivation of equation (11.42) which can be written in the following form

$$y(n) = y(n-1) + x(n) + x(n-p-1) - 2x\left(n - \frac{p}{2} - 0.5\right) \qquad (11.43)$$

The above demonstrates that a filter with a Walsh data window is very simple to calculate and - as can be seen from Fig. 11.16 - has a good filter characteristic.

The data window produced by a Walsh function of the second order is shown in Fig. 11.17a and its frequency response, which is described by the following equation, in Fig. 11.17b.

$$W_r(j\omega) = \frac{T_w}{T_i}\left[\frac{\sin\left(\frac{\omega T_w}{4}\right)}{\frac{\omega T_w}{4}}\right]\left[1 - \cos\left(\frac{\omega T_w}{4}\right)\right] \qquad (11.44)$$

The digital execution of this data window is very simple, because

$$y(n) = -\sum_{k=0}^{\frac{(p+1)}{4}-1} x(n-k) + \sum_{k=\frac{(p+1)}{4}}^{\frac{3(p+1)}{4}-1} x(n-k) - \sum_{k=\frac{3(p+1)}{4}}^{p} x(n-k) \qquad (11.45)$$

Since the calculation of equation (11.45) only requires p addition/subtraction operations, this data window can be determined very quickly indeed. Where the number of samples per data window is high, further reduction of the computation requirement can be achieved using the infinite impulse response relationship, since

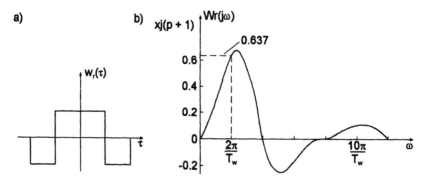

Fig. 11.17: a) data window corresponding to a second order Walsh function and b) the
corresponding amplitude spectrum

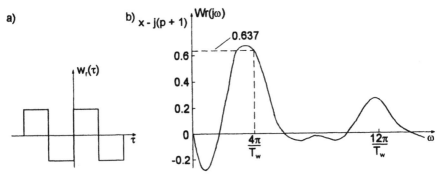

Fig. 18:   a) data window corresponding to a third order Walsh function and b) the
corresponding frequency response

$$y(n) = y(n-1) - x(n) + x(n-p-1) - 2x\left[n - \frac{(p+1)}{4}\right] -$$

$$- 2x\left[n - \frac{3(p+1)}{4}\right]$$

Discrete filters with data windows derived from Walsh functions of higher
orders can be obtained in a similar way as shown, for example, in Fig. 11.18.

11.3.3 Practical applications

In the protection and automation equipment used by power supply utilities the
filters used are of either the low-pass or bandpass type, that is filters which allow
the fundamental of the input variable to pass, but reject or attenuate all frequencies above and below ω_1.

The low-pass filters have mainly rectangular data windows, i.e. the correspond to zero order Walsh functions, because they are very simple to calculate
and exhibit relatively good filter properties (see Figures 11.7 and 11.8). Other
data windows used less frequently are, for example, triangular data windows
(Figures 11.9 and 11.10) or cosine data windows with $T_w = 0.5\ T_0$ (Fig. 11.13b).

Eliminating the fundamental from the input variable can prove problematical.
The window for such filters has a width equal to a period of the fundamental, i.e.
$T_w = T_1$, and is typically of the sine or cosine type; the frequency of the ω_0 function forming the data window is the same as the fundamental ($\omega_0 = \omega_1$). A sine
data window is chosen to reject the higher interference frequencies and a cosine

data window to reject the frequencies lower than ω_1 (e.g. exponentially decaying DC components).

In some cases concerned mainly with high-speed protection devices, a data window with a width of $T_w = T_1$ is much too wide. If the width of the data window is reduced, e.g. by half ($T_w = 0.5\,T_1$) while maintaining the relationship $\omega_0 = \omega_1$, a filter is created which has special characteristics. One negative one in the case of a cosine window is that the DC component is not removed from the measured variable (Fig. 11.13b). Thus a sine data window has to be used for widths which are a fraction of the period of the fundamental to eliminate the DC component and/or components with frequencies higher than ω_0 (Figures 11.12a and 11.13a).

As mentioned in the preceding section, the application of sine and cosine data windows involves time-consuming computations for the processor, because of the many multiplications. An alternative to overcome this problem is to use Walsh functions which do not require any multiplications. The best results with second order Walsh functions are obtained if the window width $T_w = T_1$. Such filters efficiently suppress all components with frequencies $\omega < \omega_1$, i.e. including DC components and subharmonics. Good bandpass characteristics are only obtainable with shorter data windows using first order Walsh functions (Figures 11.15 and 11.16). An optimum data window from the point of view of filter properties is the first peak of the $W_r(j\omega)$ characteristic in Fig. 11.16 corresponding to $T_w = 3T_1/4$. Reducing T_w still further to $0.5\,T_1$ only slightly diminishes the filter properties, since $\omega_1 = \pi/T_w$. Below this point or increasing T_w in relation to T_1 severely diminishes the filter properties due to the appreciable attenuation of the fundamental.

The relatively high side-bands of data windows based on first order Walsh functions are a disadvantage. For example with $T_w = 0.5\,T_1$, the attenuation of the sixth harmonic in relation to that of the fundamental is only greater by a factor of three. This can be seen from Fig. 11.15 where the amplitude spectrum $W_r(j\omega_1) = j(p + 1)0.637$ for ω_1 compared with $W_r(j\omega_6) = j(p + 1)0.21$ for ω_6.

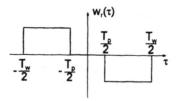

Fig. 11.19: Modified first order Walsh function

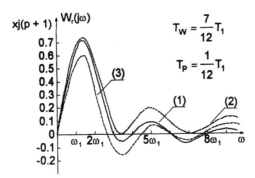

Fig. 11.20: Frequency responses of a modified first order Walsh function (1), a normal first order Walsh function (2) and a sine function (3)

The characteristics of the data window in Fig. 11.19 can be described as very good. The corresponding amplitude spectrum is given by

$$W_r(j\omega) = j\frac{T_w}{T_i}\left[\frac{\cos\left(\frac{\omega T_p}{2}\right) - \cos\left(\frac{\omega T_w}{2}\right)}{\frac{\omega T_w}{2}}\right] \tag{11.45}$$

The results for $T_p = T_1/12$ are especially good as is obvious from Fig. 11.20 which also shows for comparison the normal first order Walsh function and a sine function. The window width for all three is $T_w = 7T_1/12$. The modified Walsh

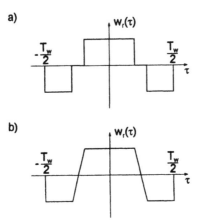

Fig. 11.21: Two examples of modified second order Walsh functions

data window is at least as efficient as the sine window, but requires much less computation.

There are many modified versions of Walsh functions which have improved frequency responses without appreciably increasing the computation requirement. Two examples for second order functions are give in Fig. 11.21.

11.4 Representation of the input variable by its orthogonal components

Algorithms frequently use the orthogonal components of the input variable. For example, for the input variable

$$x(t) = X_1 \cos(\omega_1 t - \beta) \tag{11.47}$$

the orthogonal components are described by the relationships

$$x_a = X_1 \cos(\omega_1 t - \alpha) \tag{11.48}$$

$$x_r = X_1 \cos\left(\omega_1 t - \alpha - \frac{\pi}{2}\right) \tag{11.49}$$

Thus the orthogonal components have the same amplitude as the input variable, but are shifted in relation to it by $\pi/2$ with x_a leading.

11.4.1 Application of convolution with orthogonal functions

Passing an input variable through filters with sine and cosine data windows (equations (11.34) to (11.36)) generates orthogonal signals, the one formed by convolution with the sine function leading the one formed by convolution with the cosine function by $\pi/2$ (see equations (11.30), (11.31), (11.32) and (11.34)). Thus the orthogonal signals associated with the digital signal $x(n)$ are given by the following relationships

$$x_a(n) = \frac{2}{p+1} y_s(n) \tag{11.50}$$

$$x_r(n) = \frac{2}{p+1} y_c(n) \tag{11.51}$$

in which $y_s(n)$ and $y_c(n)$ are determined by equations (11.37) and (11.38).

The orthogonal components can be similarly determined when using filters based on even and odd Walsh functions or other functions symmetrically or anti-symmetrically disposed to the center of the data window.

11.4.2 Orthogonalization by a single time delay

a) Alternative 1

An input signal determined by equation (11.47) and delayed by the time τ_1 becomes the signal

$$x(t-\tau_1) = X_1 \cos(\omega_1 t - \beta - \omega_1\tau_1) = X_1 \cos\left(\omega_1 t - \alpha - \frac{\omega_1\tau_1}{2}\right) \quad (11.52)$$

where $\alpha = \beta + \dfrac{\omega_1\tau_1}{2}$

Since

$$x(t) + x(t-\tau_1) = 2X_1 \cos(\omega_1 t - \alpha) \cos\left(\frac{\omega_1\tau_1}{2}\right) \quad (11.53)$$

and

$$x(t) - x(t-\tau_1) = -2X_1 \sin(\omega_1 t - \alpha) \sin\left(\frac{\omega_1\tau_1}{2}\right) \quad (11.54)$$

therefore

$$x_a(t) = X_1 \cos(\omega_1 t - \alpha) = \frac{x(t) + x(t-\tau_1)}{2\cos\left(\dfrac{\omega_1\tau_1}{2}\right)} \quad (11.55)$$

and

$$x_r(t) = X_1 \sin(\omega_1 t - \alpha) = \frac{x(t) - x(t-\tau_1)}{2\sin\left(\dfrac{\omega_1\tau_1}{2}\right)} \quad (11.56)$$

Any value can be chosen for the delay τ_1 which is a whole multiple of the sampling period T_i. Thus the digital form of the relationships (11.55) and (11.56) are

$$x_a(n) = \frac{x(n) + x(n-h)}{2\cos\left(\dfrac{\omega_1 h T_i}{2}\right)} \quad (11.57)$$

and

$$x_r(n) = \frac{x(n) - x(n-h)}{2\sin\left(\frac{\omega_1 h T_i}{2}\right)}$$

(11.58)

where $h = \frac{\tau_1}{T_i}$

In practice, h is chosen between 1 (a single sample) and m/4.

b) Alternative 2

An input signal determined by equation (11.47) and delayed by the time τ_2 becomes the signal

$$
\begin{aligned}
x(t - \tau_2) &= X_1 \cos(\omega_1 t - \beta - \omega_1 \tau_2) \\
&= X_1 \left[\cos(\omega_1 t - \beta)\cos\omega_1\tau_2 + \sin(\omega_1 t - \beta)\sin\omega_1\tau_2\right]
\end{aligned}
$$

(11.59)

From equations (11.47) and (11.59)

$$X_1 \sin(\omega_1 t - \beta) = \frac{x(t - \tau_2) - x(t)\cos\omega_1\tau_2}{\sin\omega_1\tau_2}$$

(11.60)

and therefore

$$x_a(t) = x(t)$$

$$x_r(t) = \frac{x(t - \tau_2) - x(t)\cos\omega_1\tau_2}{\sin\omega_1\tau_2}$$

(11.61)

From this it follows that

$$x_a(n) = x(n)$$

$$x_r(n) = \frac{x(n-h) - x(n)\cos\omega_1 h T_i}{\sin\omega_1 h T_i}$$

(11.62)

h is once again chosen between 1 and m/4.

11.4.3 Orthogonalization by a double time delay

An input signal determined by equation (11.47) and delayed by the time τ_3 and $2\tau_3$ becomes the signals

$$x(t) = X_1 \cos(\omega_1 t - \beta) = X_1 \cos(\omega_1 t - \alpha + \omega_1 \tau_3)$$

(11.63)

$$x(t - \tau_3) = X_1 \cos(\omega_1 t - \alpha) \tag{11.64}$$

$$x(t - 2\tau_3) = X_1 \cos(\omega_1 t - \alpha - \omega_1 \tau_3) \tag{11.65}$$

where $\alpha = \beta + \omega_1 \tau_3$

Considering that

$$x(t - 2\tau_3) - x(t) = 2X_1 \sin(\omega_1 t - \alpha)\sin(\omega_1 \tau_3) \tag{11.66}$$

then

$$x_a(t) = X_1 \cos(\omega_1 t - \alpha_1) = x(t - \tau_3) \tag{11.67}$$

and

$$x_r(t) = X_1 \sin(\omega_1 t - \alpha_1) = \frac{x(t - 2\tau_3) - x(t)}{2\sin(\omega_1 \tau_3)} \tag{11.68}$$

According to equation (11.62)

$$x_a(n) = x(n - h) \tag{11.69}$$

$$x_r(n) = \frac{x(n - 2h) - x(n)}{2\sin(\omega_1 h T_i)} \tag{11.70}$$

where $h = \dfrac{\tau_3}{T_i}$

11.5 Digital correlation

Digital correlation is the replacement of an input variable by a series of reversible orthogonal functions, whereby in many cases only the fundamental is determined by calculating

$$x(t) = C_1 \sin\omega_1 t + C_2 \cos\omega_1 t \tag{11.71}$$

The sine and cosine functions are orthogonal when the data window is an even multiple of half the period of ω_1. In these circumstances, the coefficients C_1 and C_2 can be written as [C.3]

$$C_1 = \frac{\displaystyle\int_{t - T_w}^{t} x(\tau)\sin(\omega_1 \tau)\, d\tau}{\displaystyle\int_{t - T_w}^{t} \sin^2(\omega_1 \tau)\, d\tau} \tag{11.72}$$

and

$$C_2 = \frac{\int\limits_{t-T_w}^{t} x(\tau)\cos(\omega_1\tau)\,d\tau}{\int\limits_{t-T_w}^{t} \cos^2(\omega_1\tau)\,d\tau} \tag{11.73}$$

In both cases, the denominator is equal to $T_w/2$ and therefore it is only necessary to determine the numerator in equations (11.72) and (11.73) to obtain the digital correlation. If the integration is executed according to Euler, then

$$c_1(n) = \frac{2}{p+1}\sum_{k=0}^{p} x(n-k)\sin\left[\omega_1(n-k)T_i\right] \tag{11.74}$$

and

$$c_2(n) = \frac{2}{p+1}\sum_{k=0}^{p} x(n-k)\cos\left[\omega_1(n-k)T_i\right] \tag{11.75}$$

The above procedures are similar to the filters with sine and cosine data windows (equations (11.34), (11.35) and (11.36)), the only difference being that the coefficients a_k are not dependent on the sequential calculation of the samples and therefore are not a function of n. As can be seen from equations (11.74) and (11.75), in the case of correlation the coefficients vary with n and therefore the algorithms for calculating the coefficients C_1 and C_2 can be treated as digital filters with coefficients which vary in relation to time.

The correlation coefficients C_1 and C_2 have the important characteristic that their value in relation to time is constant, providing the input variable only contains a single component with the frequency ω_1. Components with frequency which are a whole multiple of ω_1 have no influence on the magnitudes of C_1 and C_2.

If during calculation it is found that the coefficients C_1 and C_2 vary for n consecutive samples, it means that either

- the amplitude or the phase-angle of the input variable with the angular velocity ω_1 have varied within the prescribed time window

or

- the frequency of the fundamental has varied from its rated value ω_1

or

- the input variable contains components with angular velocities which are not whole multiples of ω_1.

Special attention must be paid to the DC component in the input variable. Providing the width of the data window is a whole multiple of the period of the fundamental (usually one period), the DC component has no influence on the values of C_1 and C_2. Should this condition not be fulfilled, which occurs for $T_w < T_l$, the coefficients are calculated with errors dependent on time.

The computation requirement to solve equations (11.74) and (11.75) appears high, but this can be greatly improved by the infinite impulse response representation of the algorithms for window widths which are whole multiples of $T_1/2$. For $T_w = kT_1/2$ and even values of k

$$c_1(n) = c_1(n-1) + \frac{2}{p+1}[x(n) - x(n-p-1)]\sin(\omega_1 nT_i) \qquad (11.76)$$

and

$$c_2(n) = c_2(n-1) + \frac{2}{p+1}[x(n) - x(n-p-1)]\cos(\omega_1 nT_i) \qquad (11.77)$$

and for odd values of k

$$c_1(n) = c_1(n-1) + \frac{2}{p+1}[x(n) + x(n-p-1)]\sin(\omega_1 nT_i) \qquad (11.78)$$

and

$$c_2(n) = c_2(n-1) + \frac{2}{p+1}[x(n) + x(n-p-1)]\cos(\omega_1 nT_i) \qquad (11.79)$$

It can be seen that consecutive values of the correlation coefficients $c_1(n)$ and $c_2(n)$ can be obtained by one multiplication per coefficient.

11.6 Symmetrical component filters

The positive, negative and zero-sequence components of three-phase sinusoidal variables are measured in order to determine the degree of their imbalance. Since an imbalance of a three-phase power system is indicative of a disturbance, the symmetrical components are often derived by protection devices in order to detect the corresponding kinds of faults (see Section 3.6).

The analogue symmetrical component filters used hitherto (see Section 3.5.1.3) produce a signal at their outputs which is proportional to one of the above components, respectively to a linear combination of more than one component.

Digital versions of the analogue symmetrical component filters could be constructed without difficulty, but it is better to take advantage of digital techniques to obtain the phase-shifts needed. Some of the principles involved are discussed below.

11.6.1 Principle based on delaying signals

The relationships between the symmetrical components x_1, x_2 and x_0 and the three phase values x_{L1}, x_{L2} and x_{L3} are given by the following expression

$$\begin{bmatrix} x_0 \\ x_1 \\ x_2 \end{bmatrix} = \frac{1}{3} \begin{bmatrix} 1 & 1 & 1 \\ 1 & a & a^2 \\ 1 & a^2 & a \end{bmatrix} \begin{bmatrix} x_{L1} \\ x_{L2} \\ x_{L3} \end{bmatrix} \tag{11.80}$$

where

x_1, x_2, x_3 = positive, negative and zero-sequence components of the phase variables x_{L1}, x_{L2} and x_{L3} with phase L1 as reference

a = operator for shifting the phase-angle of the input variable by 120°

It should be noted that a phase-shift of +120° is equivalent to lagging by 240° and a phase-shift of +240° to lagging by 120°. A lagging phase-angle, i.e. a signal delay, can be achieved digitally extremely simply, because the filter algorithms for the components are of the following form

$$3x_0(n) = x_{L1}(n) + x_{L2}(n) + x_{L3}(n)$$

$$3x_1(n) = x_{L1}(n) + x_{L2}\left(n - \frac{2m}{3}\right) + x_{L3}\left(n - \frac{m}{3}\right) \tag{11.81}$$

$$3x_2(n) = x_{L1}(n) + x_{L2}\left(n - \frac{m}{3}\right) + x_{L3}\left(n - \frac{2m}{3}\right)$$

where m is the number of samples in one period of the fundamental of the input variable.

Processing of the algorithms of (11.81) can be accelerated, since according to Fig. 11.22 the relationship $i(n - 2m/3) = -i(n - m/6)$ is fulfilled. From this results the second version of the filter algorithms

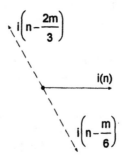

Fig. 11.22: Equivalent signal phase-shifts

$$3x_0(n) = x_{L1}(n) + x_{L2}(n) + x_{L3}(n)$$

$$3x_1(n) = x_{L1}(n) - x_{L2}\left(n - \frac{m}{6}\right) + x_{L3}\left(n - \frac{m}{3}\right) \tag{11.82}$$

$$3x_2(n) = x_{L1}(n) + x_{L2}\left(n - \frac{m}{3}\right) - x_{L3}\left(n - \frac{m}{6}\right)$$

Providing $m/3$ in equation (11.81) and $m/6$ in equation(11.82) are whole numbers, the calculation of both sets of equations is simple. Where this condition is not fulfilled, additional operations are necessary as follows.

Assuming $m/6 = k + r$ where k is a whole number and r a fraction, the following approximation applies

$$x\left(n - \frac{m}{6}\right) \cong x(n - k)(1 - r) + x(n - k - 1)r$$

Usually the value of r is such that multiplication is simple and hence the computation requirement low.

The algorithms given above are simple, fast and accurate, but only under the condition that the input variables are purely sinusoidal and have an angular velocity of ω_1. The resulting error may be considerable should this not be the case and therefore the signals derived using mixing transformers as symmetrical component filters have to be filtered additionally by infinite, respectively finite im-

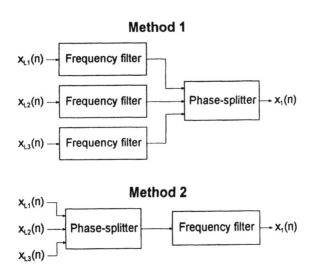

Fig. 11.23: Methods of combining frequency filters and symmetrical component filters

pulse response frequency filters (see Section 11.2). The two kinds of filters can be combined in one of the two ways illustrated in Fig. 11.23.

11.6.2 Orthogonal component filters

The symmetrical components of a three-phase system can be obtained directly using orthogonal functions. This is based on the following possibility of representing a signal which is phase-shifted by $+120°$

$$X_1 \cos\left(\omega_1 t - \alpha \pm \frac{2\pi}{3}\right)$$
$$= -X_1\left[0.5\cos(\omega_1 t - \alpha) \pm \frac{\sqrt{3}}{2}\sin(\omega_1 t - \alpha)\right]$$

(11.83)

The right-hand side of equation (11.83) is the sum of the orthogonal components. It follows from this that the positive and negative-sequence components of an AC signal are given by

$$3x_1 = (x_{L1})_a - 0.5(x_{L2})_a - \frac{\sqrt{3}}{2}(x_{L2})_r - 0.5(x_{L3})_a + \frac{\sqrt{3}}{2}(x_{L3})_r \quad (11.84)$$

$$3x_2 = (x_{L1})_a - 0.5(x_{L2})_a + \frac{\sqrt{3}}{2}(x_{L2})_r - 0.5(x_{L3})_a - \frac{\sqrt{3}}{2}(x_{L3})_r \quad (11.85)$$

where

$(x_{L1})_a, (x_{L2})_a, (x_{L3})_a$ = leading orthogonal components of the phases x_{L1}, x_{L2} and x_{L3}

$(x_{L1})_r, (x_{L2})_r, (x_{L3})_r$ = lagging orthogonal components of the phases x_{L1}, x_{L2} and x_{L3}

The orthogonal components can be determined using one of the methods described in Section 11.4.

12

Algorithms for digital protection

12.1 Introduction

The algorithms used in protection usually conform to one of the two typical structures shown in Fig. 12.1. The first (Fig. 12.1a) is based on the fact that the values of the input variables and their mutual relationships (e.g. amplitudes, frequencies and the real and apparent powers) are determined numerically. The resulting signals are processed to ascertain whether their amplitudes exceed a set limit (pick-up value), or the amplitude of one is greater or less than another, or whether the tripping condition defined by the operating characteristic of an impedance relay is fulfilled. In the subsequent parts of the structure, the logical relationships between the signals are supervised. The processes of the digital structure differ from those of the analogue structure in the respect that the input variables are measured and the internal signals processed digitally.

The second structure shown in Fig. 12.1b is used typically in analogue protection relays which do not establish the numerical values of the input variables and their relationships, but simply compare the amplitudes with a reference to decide whether the tripping conditions is fulfilled or not. Practical designs of digital protection devices are frequently a combination of the first and second structures.

a)

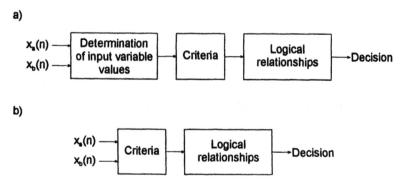

b)

Fig. 12.1:   Typical structures of measurement algorithms for protection applications

12.2 Digital measurement of the amplitude of a sinusoidal variable

Once again there are many alternative algorithms for measuring the amplitude of a sinusoidal signal which differ in their accuracy, immunity to interference, computation time requirement and computation capacity. The more important ones are described below.

12.2.1 Application of orthogonal components

First method:

The amplitude X_1 of a sinusoidal signal can be determined from the orthogonal components x_a and x_r derived with the aid of equations (11.48) and (11.49). The corresponding relationship for X_1 is

$$X_1 = \sqrt{x_a^2 + x_r^2} \tag{12.1}$$

Depending on the type of orthogonal functions determined (see Section 11.4), the ultimate measurement algorithm for equation (12.1) can take on a number of forms and also the corresponding orthogonalization parameters can play an important role.

Since the computation requirement when determining amplitude according to equation (12.1) is relatively high, simpler computation procedures are used in practice. One of these is shown in Fig. 12.2 [C.19]; the accuracy of the calculation in this case is better than 2 %.

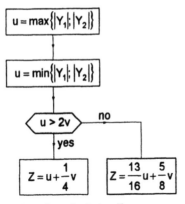

Fig. 12.2:   Simplified procedure for calculating Z

Second method:

The following relationship is obtained by delaying the orthogonal components resulting from equations (11.48) and (11.49) by τ_4

$$x_a(t)\,x_r(t - \tau_4) = 0.5X_1^2\left[-\sin(\omega_1\tau_4) + \sin(2\omega_1 t - 2\alpha - \omega_1\tau_4)\right]$$

$$x_a(t - \tau_4)\,x_r(t) = 0.5X_1^2\left[\sin(\omega_1\tau_4) + \sin(2\omega_1 t - 2\alpha - \omega_1\tau_4)\right]$$

After subtracting the two sides of these equations and rearranging

$$X_1 = \sqrt{\frac{x_a(t - \tau_4)\,x_r(t) - x_a(t)\,x_r(t - \tau_4)}{\sin(\omega_1\tau_4)}} \tag{12.2}$$

The digital form of this equation is

$$X_1(n) = \sqrt{\frac{x_a(n - 1)\,x_r(n) - x_a(n)\,x_r(n - 1)}{\sin(\omega_1 T_l)}} \tag{12.3}$$

where $l = \dfrac{\tau_4}{T_i}$

The choice is usually between $l = 1$ and $l = m/4$.

Many algorithms can be derived from equation (12.3) depending on the orthogonalization method employed which are characterized by differing degrees of immunity to signal distortion. It is generally true, however, that they are less sensitive to DC components contained in the input variable than the algorithms derived from equation (12.1). The two types of algorithms have opposite susceptibilities to HF interference.

A direct inverse relationship exists between computation speed and computation accuracy of measurement algorithms, i.e. the higher the speed, the lower the accuracy. High computation speeds are achieved, for example, when using the single and double signal delay principles. For h = 1 (see Sections 11.4.2 and 11.4.3), the measured value is available after just 2 to 4 samples, assuming that the input variable does not contain any interference. A relatively slow algorithm results from orthogonalization of the input variable by convolution with sine and cosine functions and a data window width equal to one period of the fundamental which with a settling time of T_1 does, however, have the advantage of being immune to signal distortion.

12.2.2 Utilization of the correlation coefficients

The correlations of sine and cosine functions needed to calculate the coefficients C_1 and C_2 were described in Section 11.2 on the basis of the approximation

$$x(t) \cong C_1 \sin\omega_1 t + C_2 \cos\omega_1 t$$

For this case, the amplitude of the fundamental can be calculated as follows

$$X_1 = \sqrt{C_1^2 + C_2^2} \tag{12.4}$$

The discrete form of X1 can be determined when using algorithms (e.g. equation (11.76) or (11.77)) to calculate the coefficients C_1 and C_2 or the infinite impulse response relationships of equations (11.78) and (11.79) or (11.80) and (11.81) with the aid of

$$X_1(n) = \sqrt{C_1^2(n) + C_2^2(n)} \tag{12.5}$$

12.2.3 Averaging the absolute values of signals

It is known that taken over a whole or half a period, the mean values of rectified sinusoidal signals are proportional to the amplitude X_1, but the factor of proportionality is independent of the location of the data window. This is illustrated in Fig. 12.3 and can be expressed as follows

$$\int_{t-T_1}^{t} |x(\tau)| d\tau = \frac{4X_1}{\omega_1} \tag{12.6}$$

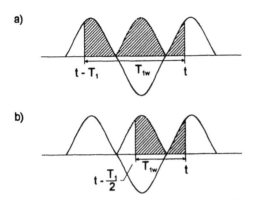

Fig. 12.3:   Amplitude measurement by averaging the absolute value
             a) one-period data window,   b) half-period data window

$$\int_{t-\frac{T_1}{2}}^{t} |x(\tau)| d\tau = \frac{2X_1}{\omega_1} \qquad (12.7)$$

The following is obtained by applying the simplest integration procedure according to Euler

$$X_1(n) = \frac{\pi}{2(p+1)} \sum_{k=0}^{p} |x(n-k)| \qquad (12.8)$$

If $T_{1w} = T_1$ (equation (12.6) and Fig. 12.3a), $p + 1 = m$, but for $T_w = 0.5\, T_1$ (equation (12.7) and Fig. 12.3b), $p + 1 = m/2$.

The calculation of equation (12.8) is simple, because only the absolute value is needed and the addition has to be performed to determine the variable proportional to X_1 using the factor of proportionality $\pi/2(p + 1)$. Equation (12.8) can be further simplified by using the infinite impulse response relationship

$$X_1(n) = X_1(n-1) + \frac{\pi}{2(p+1)} \left[|x(n)| - |x(n-p-1)| \right] \qquad (12.9)$$

Care must be taken when using this equation, because a change in the values for $X_1(n - 1)$ stored in the register due to interference can cause irreversible errors and a program crash.

While algorithm (12.8) is not only economical from the computation point of view and insensitive to the influence of frequencies higher than ω_1, it is adversely influenced by any DC components in the input variable. Thus where DC components are to be expected, they must be eliminated by filtering beforehand. Figure

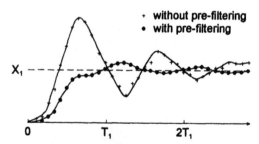

Fig. 12.4: Amplitudes x_1 for the input variable of (12.10) calculated using equation
(12.8)

$$b_0 = 1, \qquad b_k = 0.25, \qquad \alpha_1 = 0, \qquad \frac{\omega_k}{\omega_1} = 5.5$$

12.4 shows the results when determining the absolute value for half a period, i.e.
for $T_w = 0.5\ T_1$.

Pre-filtering was performed by convolution with a first order Walsh function.
The calculations assume that the input variable is of the form

$$x(t) = X_1 \left[\begin{array}{l} \cos(\omega t - \alpha_1) - b_0 \cos\alpha_1 \exp\left(-\dfrac{t}{T_a}\right) + \\[2mm] + b_k \sin(\omega_k t - \alpha_k) \end{array} \right] \tag{12.10}$$

a) Averaging using a single phase-shift

The averaging procedure can be modified by including not only the input variable
$x(t)$ itself, but also its image $x(t - \tau_1)$ which is delayed by $T_1/4$, i.e.

$$x(t) = X_1 \sin(\omega_1 t - \alpha_1)$$

and

$$x(t - \tau_1) = X_1 \sin[\omega_1(t - \tau_1) - \alpha_1]$$

where $\tau_1 = \dfrac{\pi}{2\omega_1}$

Taking only the greater instantaneous value of the two signals, the resultant is
the bold curve in Fig. 12.5 of the function $x^*(t)$ given by

$$x^*(t) = \max\{|x(t)|; |x(t - \tau_1)|\}$$

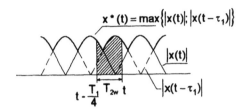

Fig. 12.5:   Averaging based on a single phase-shift

The following now applies

$$\int_{t-\frac{T_1}{4}}^{t} x^*(\tau)d\tau = \sqrt{2}\,\frac{X_1}{\omega_1}$$

(12.11)

Integrating according to Euler, the discrete form of equation (12.11) becomes

$$x^*(n) = \max\left\{|x(n)|;\ \left|x\left(n-\frac{m}{4}\right)\right|\right\}$$

(12.12)

$$X_1(n) = \frac{\sqrt{2}\pi}{m}\sum_{k=0}^{\frac{m}{4}-1} x^*(n-k)$$

(12.13)

Note that m is the number of samples per period T_1. Only a low computation requirement is needed to calculate the variable proportional to $X_1(n)$ using the factor of proportionality $\sqrt{2}\pi/m$, because there are only one comparison and $[(m/4)-1]$ addition operations (expression (12.13)) to be made. Theoretically, the number of additions can be reduced by the infinite impulse response version of the algorithm, but the practical improvement is so small that this procedure is seldom used.

b) Averaging using a double phase-shift

By phase-shifting the input variable by $\pi/3$ and $2\pi/3$, the same result is achieved as if a single-phase variable were replaced by a three-phase variable, since

$$x(t) = X_1 \sin(\omega_1 t - \alpha_1)$$

$$x(t - \tau_2) = -X_1 \sin\left(\omega_1 t - \alpha_1 - \frac{4\pi}{3}\right)$$

$$x(t - 2\tau_2) = X_1 \sin\left(\omega_1 t - \alpha_1 - \frac{2\pi}{3}\right)$$

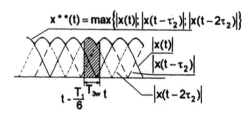

Fig. 12.6:   Averaging based on a double phase-shift

where $\tau_2 = \dfrac{\pi}{3\omega_1}$

Taking only the greater instantaneous value of the three signals, the resultant is the bold curve in Fig. 12.6 of the function $x^{**}(t)$ given by

$$x^{**}(t) = \max\left\{|x(t)|;\, |x(t - \tau_2)|;\, |x(t - 2\tau_2)|\right\}$$

It can now be demonstrated that

$$\int_{t - \frac{T_1}{6}}^{t} x^{**}(\tau)d\tau = \frac{X_1}{\omega_1} \tag{12.14}$$

Integrating according to Euler, the discrete form of equation (12.14) becomes

$$x^{**}(n) = \max\left\{|x(n)|;\, \left|x\left(n - \frac{m}{6}\right)\right|;\, \left|x\left(n - \frac{m}{3}\right)\right|\right\} \tag{12.15}$$

$$X_1(n) = \frac{2\pi}{m}\sum_{k=0}^{\frac{m}{6}-1} x^{**}(n-k) \tag{12.16}$$

Once again only a low computation requirement is needed for the final algorithm, because there are only two comparisons and $\left[(m/6) - 1\right]$ addition operations to be made.

c) Determining the maxima of phase-shifted signals

The maximum during a half-period of a sinusoidal variable is easily obtained by calculating the maximum of the absolute value.

$$X_1 = \max\left\{|x(\tau)|\right\}\Big|_{\tau = t - \frac{T_1}{2}}^{t} \tag{12.17}$$

In the discrete form, the result is the maximum absolute value of the last $m/2$ samples of the input variable x, i.e.

$$X_1(n) = \max\left\{|x(n-k)|\right\}\Big|_{k=0}^{\frac{m}{2}-1} \tag{12.18}$$

This algorithm has the advantage of only requiring $\left[(m/2)-1\right]$ comparisons of the absolute values. An additional advantage is that the result equals the signal amplitude, i.e. the factor of proportionality is 1. A disadvantage is its relatively high sensitivity to HF components in the signal. Also at low sampling frequencies, the error ΔX_1 can become appreciable. The maximum error is given by

$$\Delta X_1 = X_1\left[1 - \cos\frac{\omega_1 T_1}{2}\right]$$

However, the results achieved with this algorithm are only dependent to a small extent on frequency fluctuations of the fundamental.

12.2.4 Application recommendations

In view of the many procedures for calculating the amplitude of sinusoidal input variables which have been described, the real problem is choosing the right one for the application in hand. This, however, at best can only be a compromise taking the following factors into account

- required time after which a steady-state result has to be available
- required steady-state and transient accuracies, especially the calculation overshoot following a dynamic change of the input variable
- expected proportions of frequencies other than ω_1 and their type (DC component, damped oscillation, harmonic etc.)
- computation requirement

The main interest in protection is the measurement of the fundamentals of current and voltage. These input variables have to be processed separately.

In a voltage signal, the following components with frequencies other than the fundamental must be expected

a) DC components generated by some phase faults close to the relay location. Their initial value is relatively low, usually only a fraction of the rated voltage, and they decay exponentially at a time constant of 0.01 to 0.1 seconds.

b) HF oscillations following dynamic changes such as faults or switching operations. The frequencies of these components are normally so high that they can be removed without difficulty by analogue filters (see Section 10.2). Only seldom is the frequency less than 600 Hz, e.g. a fault at the end of a very long

HV or EHV transmission line. The amplitudes reached by HF interference can be very high, in extreme cases equaling the amplitude of the fundamental; they decay, however, rapidly at time constants of up to 0.1 seconds.

c) harmonics caused by non-linearities of system plant impedances. The odd harmonics often lie in the admittance range of analogue filters and have amplitudes of a few percent of the fundamental.

It can be concluded from the above that protection devices which only measure the voltage amplitude and not its relationship to the current do not require any digital pre-filtering. The use of algorithms which are insensitive to HF components and only involve little computation is recommended. The algorithms described by equations (12.8) and (12.12) belong to this category.

Conditioning the signals by passing them through digital bandpass filters prior to processing is essential for protection devices which evaluate the relationship between voltage and current, e.g. distance or phase-angle relays. Filters based on the convolution of the modified first order Walsh function (see Fig. 11.19) are especially recommended. In this case, the best results are obtained when using algorithm (12.8), (12.12) or (12.15) to calculate the voltage amplitude.

Basically, the current signals used in protection devices contain the same undesired components as the voltage signals. The only difference is in their magnitude

a) DC components can have an initial value equal to the amplitude of the AC fault current and decay at time constants between 0.05 and 0.8 seconds. They can thus bear a decisive influence on the results of computations.

b) HF oscillations normally have frequencies which are seldom less than 600 Hz and amplitudes seldom greater than 10 % of the fundamental.

c) The amplitudes of harmonics in a fault current seldom exceed 10 % of that of the fundamental. They can, however, increase disproportionately should c.t's saturate. High proportions of odd harmonics (possibly with amplitudes higher than the fundamental) are to be expected in ground fault currents in ungrounded systems and systems with Petersen coils.

The conclusion in this case is that digital filtering is superfluous when measuring currents under steady-state conditions (e.g. time-overcurrent relays with long time delays). All the algorithms presented in Section 12.2, i.e. the algorithms described by equations (12.8), (12.12) and (12.15), fulfil the requirements in this respect and are to be recommended, because they involve the least computation.

Digital filtering is unavoidable where a fault current has to be measured and evaluated by an overcurrent device immediately a fault occurs. The principal objective is the removal of the DC component which is best achieved by filters based on the convolution of a modified first order Walsh function with data windows $T_w = 0.5\ T_1$. If the protection device has to have a short time delay of a few hundred milliseconds, a data window in the form of a second order Walsh function or its modified variant (see Fig. 11.21) with $T_w = T_1$ is a very efficient DC component suppressor.

Following conditioning in a digital filter, the appropriate algorithms can be used to measure the amplitude of a current signal. The algorithms for this purpose with the highest speed, least computation requirement and lowest computation overshoot are equations (12.12) and (12.15).

Example 12.1

The input variable is given by equation (12.8) with $b_1 = 1$ and $b_2 = 0.25$. The sampling rate f_i is 600 Hz, i.e. there are 12 samples per period. Two different types of filters and measurement procedures are used to determine the amplitude of the fundamental.

Alternative 1: Pre-filtering is accomplished by convolution of the modified first order Walsh function with $T_w = 7T_1/12$ and $T_p = T_1/12$ (neglects the middle sample). The amplitude is determined by averaging and double phase-shift, i.e. using algorithms (12.15) and (12.16). The following are thus obtained

$$y(n) = x(n) + x(n-1) + x(n-2) - x(n-4) - x(n-5) - x(n-6)$$

$$y**(n) = \max\{|y(n)|;\ |y(n-2)|;\ |y(n-4)|\}$$

$$8.98X_1 = y**(n) + y**(n-1)$$

The curve corresponding to the results is shown in Fig. 12.7. Considering the appreciable distortion of the input variable, the result can be described as very accurate, especially as the overshoot even for long DC component time constants Ta does not exceed 12 %.

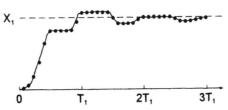

Fig. 12.7: Calculating the amplitude X_1 of the signal according to equation (12.8)
Alternative 1
$b_0 = 1,\quad \alpha_1 = 0,\quad T_a = 0.02,\quad b_k = 0$

Alternative 2: Pre-filtering is accomplished by convolution of the modified second order Walsh function (Fig. 11.21b) with $T_w = T_1$. The amplitude is determined by averaging and single phase-shift (equations (12.12) and (12.13)). The following are thus obtained

$$y(n) = -x(n) - x(n-1) + x(n-4) + x(n-5) + x(n-6) +$$
$$+ x(n-7) - x(n-10) - x(n-11) +$$
$$+ 0.5[x(n-3) + x(n-8) - x(n-2) - x(n-9)]$$

$$y^*(n) = \max\{|y(n)|; |y(n-3)|\}$$

$$20.25X_1(n) = y^*(n) + y^*(n-1) + y^*(n-2)$$

The curve corresponding to the resulting amplitude for X_1 is shown in Fig. 12.8. It can be seen that the final steady-state value is reached a little slower than Alternative 1, but the accuracy is much higher and the overshoot does not exceed 3 %.

Both the above alternatives can be recommended for practical applications, especially as the computation requirement is low.

The neutral current is also measured in addition to the phase currents mainly for ground fault protection purposes in ungrounded systems and systems with Petersen coils. Since this kind of protection normally always operates after a time delay, it is adequate to simply reduce the influence of the odd harmonics on the computation accuracy. The best filter properties are achieved by convolution of sine and cosine functions with data window widths of $T_w = T_1$. This procedure does, however, involve a relatively high computation requirement, although one of the faster averaging algorithms could be used to determine the amplitude of an input variable which has already been conditioned. An alternative is the correlation of sine and cosine functions with a data window of $T_w = T_1$, which needs less computation for the infinite impulse relationships, and then the amplitude of

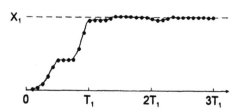

Fig. 12.8:   Calculating the amplitude X_1 of the signal according to equation (12.8) Alternative 2

$b_0 = 1, \quad \alpha_1 = 0, \quad T_a = 0.02, \quad b_k = 0$

the input variable can be calculated using algorithm (12.5). Both the recommended alternatives eliminate the influence of HF oscillations on the calculated result for I_1.

12.3 Measurement of real and apparent components

Assuming two sinusoidal input variables, one a current and the other a voltage, with curves given by the following relationships

$$i(t) = I_1 \sin(\omega_1 t - \alpha_1 + \varphi)$$
$$u(t) = U_1 \sin(\omega_1 t - \alpha_1)$$

it is necessary depending on the application to determine the component of one in relation to the other which is either in phase or shifted by $\pi/2$, i.e. either the real or apparent components have to be measured. For example, for a sinusoidal current these are

$$I_{re} = I_1 \cos \varphi$$

$$I_{im} = I_1 \sin \varphi$$

(12.19)

where

I_{re} = real component of the current
I_{im} = apparent component of the current

The following Sections describe different discrete procedures for determining the corresponding component of the input variable.

12.3.1 Determining the component at the zero-crossing of the input variable

One of the simplest and most frequently applied method of determining the apparent component of a current variable in analogue systems is to measure its value at the zero-crossing of the associated voltage (Fig. 12.9). From the above relationship, the current at time t'_1 (voltage zero for a positive-going signal) the current is given by

$$i(t'_1) = I_1 \sin \varphi$$

The corresponding relationship for the time t'_2 (voltage zero for a negative-going signal) is

$$i(t'_2) = -I_1 \sin \varphi$$

If instead of the signal i(t), its value at time $\tau = T_1/4$ is inserted, i.e.

$$i(t-\tau) = I_1 \sin\left(\omega_1 t - \alpha_1 + \varphi - \frac{\pi}{2}\right)$$

The current at time t'_1 is

$$i(t'_1-\tau) = -I_1 \cos\varphi$$

and at time t'_2

$$i(t'_2-\tau) = I_1 \cos\varphi$$

The digital algorithm for calculating the real and apparent components of signals can now be derived from these equations. This is made somewhat difficult, however, by the fact that the zero-crossing of the voltage is only approximately known, i.e. that it took place between samples (n) and (n – 1) (Fig. 12.10). It is possible to get closer to the instant of the zero-crossing if it is assumed that the voltage varies approximately linearly between the two samples with results of opposite polarity. In this case, the following is true

$$t_2{}^* = \frac{t_2}{T_i} = \frac{u(n)}{u(n)-u(n-1)} \tag{12.20}$$

Assuming also that the current varies linearly during the same period, then

$$i_x = I_1 \sin\varphi = i(n) - [i(n) - i(n-1)]t_2{}^* \tag{12.21}$$

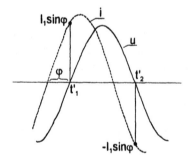

Fig. 12.9:   Relationship between voltage and current when determining the apparent component

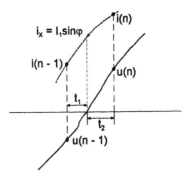

Fig. 12.10:   Close determination of the zero-crossing of an input variable

Considering that the apparent component has to be calculated for positive and negative-going zero-crossings of the voltage waveform, the final form of the equation for the algorithm becomes

$$I_1 \sin\varphi = \left\{ i(n) - \left[i(n) - i(n-1) \right] t_2 \, ^* \right\} SGN\left[u(n) - u(n-1) \right] \qquad (12.22)$$

where SGN [z] is equal to 1, −1 or zero according to whether the argument of z is positive, negative or zero.

The execution of algorithm (12.22) involves one division and one multiplication which take time, but it is only executed twice per period after each zero-crossing of the voltage and not after every sample.

The real component is determined in a similar manner, the phase-shift of the current being $T_1/4$. Thus

$$I_1 \cos\varphi = -\left\{ i\left(n - \frac{m}{4}\right) - \left[i\left(n - \frac{m}{4}\right) - i\left(n - 1 - \frac{m}{4}\right) \right] t_2 \, ^* \right\} \times$$
$$\times SGN\left[u(n) - u(n-1) \right] \qquad (12.23)$$

A division and two multiplications are therefore necessary to determine the real and apparent components.

A serious shortcoming of this algorithm is the considerable error which occurs if the input variable is distorted. In order to achieve satisfactory results in this circumstance as well, the input variables have to pass through digital band-pass filters. It is also necessary to check that the error resulting from equation (12.20) is not too large.

12.3.2 Determination of the components by averaging

Apart from the voltage and current waveforms, Fig. 12.11 also shows the curve of the function

$$i^*(t) = i(t) \, SGN[u(t)]$$

It can be simply proven that the desired components can be calculated using the following equations independently of the time t.

$$I_1 \cos\varphi = \frac{\omega_1}{2} \int_{t-\frac{T_1}{2}}^{t} i^*(\tau)d\tau$$

$$I_1 \sin\varphi = \frac{\omega_1}{2} \int_{t-\frac{T_1}{2}}^{t} i^*(\tau - \tau_1)d\tau \qquad\qquad (12.24)$$

where $\tau_1 = \dfrac{T_1}{4}$

The digital derivation of equation (12.24) is also very simple, since either Euler's principle or — for greater accuracy — the trapezium rule can be applied. Nevertheless, there is a considerable error due to the imprecise knowledge of the zero-crossing of the voltage. For this reason, a correction is made for the region in which the voltage changes its sign. The algorithm relies on the fact that a new variable y(n) is calculated for every new current and voltage sample which corresponds to the integral of the current i(t) during the last sample interval T_i. Integrating according to the trapezium rule

$$|u(n) + u(n-1)| \ge |u(n) - u(n-1)|$$

i.e. if the sign of the voltage has not changed

$$y(n) = 0.5[i(n) + i(n-1)] \, SGN[u(n) + u(n-1)] \qquad\qquad (12.25)$$

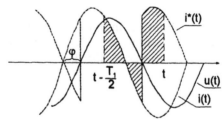

Fig. 12.11: Determination of the real component by averaging

Should, however,

$$\left|u(n)+u(n-1)\right| < \left|u(n)-u(n-1)\right|$$

i.e. if the voltage u(t) changed its sign during the last sample interval, then

$$y(n) = 0.5\left\{t_2 {}^*\left[2i_x + i(n) - i(n-1)\right] - \left[i_x + i(n-1)\right]\right\} \times$$
$$\times SGN[u(n)]$$
(12.26)

where $t_2{}^*$ and i_x are given by equation (12.20) and (12.21) respectively.

The real component of the current can now be determined from the following relationship

$$I_1 \cos \varphi = \frac{\pi}{m} \sum_{k=0}^{\frac{m}{2}-1} y(n-k)$$
(12.27)

The apparent component of the current is determined in a similar manner using current samples which are delayed by m/4, i.e. according to equation (12.27). The following substitutions are necessary in order to calculate y(n) using equations (12.25) and (12.26)

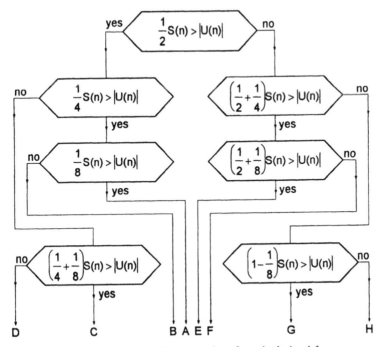

Fig. 12.12: Algorithms for the simplified procedure for calculating $t_2{}^*$

Table 12.1 Results of the calculations for t_2^* in Fig. 12.12

	A	B	C	D	E	F	G	H
t_2^*	$\dfrac{1}{16}$	$\dfrac{1}{4}-\dfrac{1}{16}$	$\dfrac{1}{4}+\dfrac{1}{16}$	$\dfrac{1}{2}-\dfrac{1}{16}$	$\dfrac{1}{2}+\dfrac{1}{16}$	$\dfrac{3}{4}-\dfrac{1}{16}$	$\dfrac{3}{4}+\dfrac{1}{16}$	$1-\dfrac{1}{16}$

$i\left(n-\dfrac{m}{4}\right)$ in place of $i(n)$

$i\left(n-\dfrac{m}{4}-1\right)$ in place of $i(n-1)$

The fact that the absolute value of the current i_x equals the absolute value of the apparent current can be used to prove the calculations, i.e.

$$|i_x| \cong |I_1 \sin\varphi|$$

There is a simple possibility of estimating the value of t_2^*; if $u(t)$ changes its sign between the n th. and the $(n-1)$ th. sample, then

$$s(n) = |u(n) - u(n-1)|$$

This is followed by the series of calculations for T_2 in the sequence shown in Fig. 12.12, for which the results A to H are given in Table 12.1.

The above procedure has two basic advantages

1. The estimated value for t_2 is obtained by three comparisons.
2. The value t_2 is given as the sum of numbers to the power of 2 which simplifies the multiplication by t_2 and only requires little computation.

The simplified determination of the zero-crossing of the voltage causes an error which in the case of the above procedure is in the range

$$\frac{\Delta t_2}{T_i} = \pm\frac{1}{16}$$

12.4 Digital measurement of power

12.4.1 Power measurement by correlation of current and voltage

The product of the input variables

$$i(t) = I_1 \cos(\omega_1 t - \alpha_1 + \varphi)$$
$$u(t) = U_1 \cos(\omega_1 t - \alpha_1)$$

(12.28)

is

$$i(t)\, u(t) = \frac{1}{2} U_1\, I_1 \left[\cos\varphi + \cos(2\omega_1 t - 2\alpha_1 + \varphi)\right]$$

Integrating the last equation between the limits $(t - T_1/2)$ and t

$$P = \frac{1}{2} U_1\, I_1 \cos\varphi = \frac{2}{T_1} \int_{t-\frac{T_1}{2}}^{t} u(\tau)\, i(\tau)\, d\tau$$

(12.29)

The apparent power is obtained by delaying the current signal by $\tau_1 = T_1/4$ so that

$$Q = \frac{1}{2} U_1\, I_1 \sin(-\varphi) = \frac{2}{T_1} \int_{t-\frac{T_1}{2}}^{t} u(\tau)\, i(\tau - \tau_1)\, d\tau$$

(12.30)

Alternatively the voltage can by delayed by τ_1 instead of the current, i.e.

$$Q = \frac{1}{2} U_1\, I_1 \sin(-\varphi) = \frac{2}{T_1} \int_{t-\frac{T_1}{2}}^{t} u(\tau - \tau_1)\, i(\tau)\, d\tau$$

(12.31)

Integrating equations (12.28), (12.30) and (12.31) according to Euler produces the discrete algorithms

$$P = \frac{2}{m} \sum_{k=0}^{\frac{m}{2}-1} u(n-k)\, i(n-k)$$

$$Q = -\frac{2}{m} \sum_{k=0}^{\frac{m}{2}-1} u(n-k)\, i\left(n - \frac{m}{4} - k\right)$$

(12.32)

$$Q = \frac{2}{m} \sum_{k=0}^{\frac{m}{2}-1} u\left(n - \frac{m}{4} - k\right) i(n-k)$$

These equations are the obvious algorithms for calculating real and apparent power.

12.4.2 Utilization of orthogonal current and voltage components

a) Without signal delay

If the currents and voltages are given as orthogonal components, then

$$u_a(t) = U_1 \cos(\omega_1 t - \alpha_1)$$
$$u_r(t) = U_1 \sin(\omega_1 t - \alpha_1)$$
$$i_a(t) = I_1 \cos(\omega_1 t - \alpha_1 + \varphi)$$ (12.33)
$$i_r(t) = I_1 \sin(\omega_1 t - \alpha_1 + \varphi)$$

From the relationships between the trigonometrical functions and their products

$$P = 0.5\, U_1\, I_1 \cos\varphi = 0.5\left[u_a(t)\, i_a(t) + u_r(t)\, i_r(t)\right]$$
$$Q = -0.5\, U_1\, I_1 \sin\varphi = 0.5\left[u_r(t)\, i_a(t) - u_a(t)\, i_r(t)\right]$$ (12.34)

The procedures for determining the orthogonal components were described in Section 11.4. Each one of them can be used for calculating the value of the power. Different algorithms can be derived according to the orthogonalization method and the number of samples h which delay the signal. Those with h = 1 are the fastest, one of which is given by the following relationships

$$P = 0.5 \frac{\left[i(n) + i(n-1)\right]\left[u(n) + u(n-1)\right]}{4\cos^2\left(\dfrac{\omega_1 T_i}{2}\right)} +$$

$$+ 0.5 \frac{\left[i(n) - i(n-1)\right]\left[u(n) - u(n-1)\right]}{4\sin^2\left(\dfrac{\omega_1 T_i}{2}\right)}$$ (12.35)

$$Q = 0.5 \frac{[i(n)+i(n-1)][u(n)-u(n-1)]}{2\sin(\omega_1 T_1)} -$$

$$- 0.5 \frac{[i(n)-i(n-1)][u(n)+u(n-1)]}{2\sin\left(\frac{\omega_1 T_1}{2}\right)}$$

(12.36)

These algorithms enable the calculations to be completed two sample periods after the change in the signals, but the signals have to be filtered first.

The orthogonalization procedure with $h = m/4$ discussed in Section 11.4.3 is used more frequently in practice. For this case

$$P = 0.5\left[i(n)\,u(n) + i\left(n-\frac{m}{4}\right)u\left(n-\frac{m}{4}\right)\right]$$

(12.37)

$$Q = 0.5\left[i(n)\,u\left(n-\frac{m}{4}\right) - i\left(n-\frac{m}{4}\right)u(n)\right]$$

(12.38)

The influence of HF interference on measuring accuracy of these algorithms is only slight.

The slowest orthogonal method, but the one with the least sensitivity to HF interference, is based on the convolution of sine and cosine functions or first or second order Walsh functions with a data window equal to T_1. The following applies to the convolution of sine and cosine functions

$$P = \frac{2}{(p+1)^2}\left[i_s(n)\,u_s(n) + i_c(n)\,u_c(n)\right]$$

(12.39)

$$Q = \frac{2}{(p+1)^2}\left[i_s(n)\,u_c(n) - i_c(n)\,u_s(n)\right]$$

(12.40)

where

$i_s(n)$, $u_s(n)$ = samples resulting from convolution of the currents, respectively voltages with the sine function

$i_c(n)$, $u_c(n)$ = samples resulting from convolution of the currents, respectively voltages with the cosine function

$(p + 1)$ = number of samples in the data window for the convolution operation (i.e. filtering by convolution) according to equations (11.37) and (11.38)

Similar equations result from orthogonalization of the input variables by convolution of first and second order Walsh functions.

b) With signal delay

The following relationships apply if the orthogonal components are given by equations (12.33)

$$u_a(t - \tau_4)\, i_r(t) - u_a(t)\, i_r(t - \tau_4) = U_1\, I_1 \cos\varphi\, \sin\omega_1\tau_4$$

$$u_a(t)\, i_a(t - \tau_4) - u_a(t - \tau_4)\, i_a(t) = U_1\, I_1 \sin\varphi\, \sin\omega_1\tau_4$$

(12.41)

and the discrete relationships of these are

$$P = \frac{\left[u_a(n-1)\, i_r(n) - u_a(n)\, i_r(n-1) \right]}{2 \sin(\omega_1 T_i\, I)}$$

(12.42)

$$Q = \frac{\left[u_a(n)\, i_a(n-1) - u_a(n-1)\, i_a(n) \right]}{2 \sin(\omega_1 T_i\, I)}$$

(12.43)

where $T_i \dfrac{\tau_4}{I}$

The factor I is usually chosen between 1 and $m/4$.

12.4.3 Utilization of correlation coefficients

The currents and voltages described by the equations (12.41) can be written in the following form

$$i(t) = C_{1i} \sin\omega_1 t + C_{2i} \cos\omega_1 t$$

$$u(t) = C_{1u} \sin\omega_1 t + C_{2u} \cos\omega_1 t$$

where

$$C_{1i} = I_1 \sin(\alpha_1 - \varphi)$$

$$C_{1u} = U_1 \sin\alpha_1$$

$$C_{2i} = I_1 \cos(\alpha_1 - \varphi)$$

$$C_{2u} = U_1 \cos\alpha_1$$

It follows that

$$P = 0.5\,U_1\,I_1\cos\varphi = 0.5\left(C_{2u}\,C_{2i} + C_{1u}\,C_{1i}\right) \qquad (12.44)$$

$$Q = -0.5\,U_1\,I_1\sin\varphi = 0.5\left(C_{1u}\,C_{2i} - C_{2u}\,C_{1i}\right) \qquad (12.45)$$

The correlation of the input variables was defined in Section 11.2 with the aid of sine and cosine functions. Thus all the coefficients needed to solve equations (12.44) and (12.45) can be determined using algorithms (11.76) and (11.77) or their associated infinite impulse response relationships (11.78) and (11.79).

12.5 Digital distance measurement

By far the greater part of all the publications on the subject of selective digital protection are concerned with the determination of the distance between the relay location and faults on HV overhead lines. This complex of problems is so dominant that even the equations for the discrete measurement of the amplitudes of sinusoidal input variables or of real and apparent power were mere "byproducts" during the development of the algorithms for calculating reactance and resistance.

As is the case with analogue protection devices, the operating principle of most digital distance protection systems is based on the determination of whether the values of resistance and reactance (or resistance and inductance) measured at the relay location lie inside the operating characteristic, thus indicating a fault in the zone of protection (Fig. 12.13).

The majority of digital distance relays perform their task by first calculating the resistance R_S and the reactance $X_S = \omega_1 L_S$ from the given input variables for the current $i(t)$ and voltage $u(t)$ and then determining whether the values obtained lie within the operating characteristic in the R/X plane.

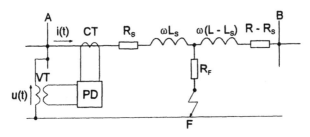

Fig. 12.13:   Distance protection measuring principle
PD = protection device, R_F = fault resistance

There are many factors which influence the accuracy of the values determined for the resistance and reactance of a fault on an overhead line. Amongst these are

- distortion of the current and voltage curves by decaying DC components and HF oscillations
- inductive coupling and insulation breakdown between the conductors of the faulted line, respectively between the faulted and healthy circuits of a double-circuit line
- fault resistance (R_F in Fig. 12.13)
- c.t. and v.t. errors, primarily saturation phenomena caused by decaying DC components

The derivations of the algorithms for calculating R_S and X_S generally only take account of the first of the above factors, because it is assumed that the mutual impedance between conductors can be compensated with the aid of the zero-sequence component, respectively with the aid of the currents of healthy lines. It is also assumed that the amplitude and phase errors of c.t's and v.t's and the error due to the fault resistance R_F are relatively small compared with the reactance X_S and may therefore be neglected. As is explained later, these assumptions are not permissible in all cases. The basic procedures for determining R_S and X_S are the subject of the following Sections.

12.5.1 Utilization of real and apparent power values

The various methods of measuring real and apparent power on the basis of discrete currents and voltages were described in Section 12.4. All the methods assume that the input variables only contain the fundamental. Once the values of these powers are known, R_S and X_S can be calculated from the following relationships

$$R_S = \frac{2P}{I_1^2}$$

$$X_S = \frac{2Q}{I_1^2}$$

(12.46)

where
P, Q = real and apparent powers
I_1 = amplitude of the fundamental i(t)

On the basis of the various procedures for calculating P and Q (see Section 12.4) and the amplitudes of the input variables (see Section 12.2), a variety of algorithms for calculating R_S and X_S can be derived from the equations (12.46), which differ in their computation times and requirement and their accuracy when processing distorted signals. The speed and accuracy of distance measurement using X_S and R_S depends on the following factors

- pre-filtering of input variables
- the method of orthogonalization in use
- the method employed for measuring the amplitudes of the current I_1 and the powers P and Q

Digital pre-filtering is essential, if no orthogonalization by convolution is performed. Filters having a finite impulse response and data windows in the form of first order Walsh functions with $T_w = 0.5\ T_1$ are generally adequate.

The fastest algorithms for determining X_S and R_S are obtained by orthogonalization based on delaying signals with low values of the delay factor h, but the smaller h, the larger the calculation error for input variables with frequencies higher than ω_1. Orthogonalization by convolution (Section 11.4.1) achieves greater accuracy for $T_w = T_1$, the susceptibility to DC components grows, however, for shorter data windows.

The computing times and accuracies of the algorithms used to calculate I_1, P and Q also vary in the following respects

- The algorithms for averaging input variables (equations (12.8) and (12.32)) are relatively slow, but insensitive to interference with angular velocities $\omega > \omega_1$.
- Equations (12.1) and (12.34) are very fast, but there accuracy is poor in the presence of HF interference.
- The algorithms for determining X_S and R_S by correlation (equations (12.4), (12.44) and (12.45)) are very accurate, but slow.
- The utilization of orthogonal components (equations (12.3), 12.42) and (12.43)) achieves short computation times (especially if I is small) and the influence of DC components on accuracy is low, but the susceptibility to HF interference is high.

12.5.2 Utilization of real and apparent components of current and voltage

a) Determination of resistance and reactance

The resistance and reactance of a short-circuited length of line between relay and fault locations can be defined in terms of the real and apparent components of the voltage signal as follows

$$R_s = \frac{U_1 \cos\varphi}{I_1}$$

$$X_s = \frac{U_1 \sin(-\varphi)}{I_1}$$

(12.47)

where φ is the phase-angle by which the fundamental of the current leads the fundamental of the voltage.

The procedure for determining the real and apparent components of the current was described in Section 12.3.1 and the same procedure can be applied to determine the corresponding voltage components. The equations for calculating the resistance and reactance after every zero-crossing $i(t)$ of the current waveform can be obtained by correspondingly rearranging equations (12.22) and (12.23). Their form also depends on the algorithm used to calculate the amplitude of the current I_1. The best results are obtained with what are referred to as averaging algorithms (see equations (12.8) or (12.12) and (12.13) or (12.15) and (12.16) in Section 12.2.3). The algorithm (12.18) which is recommended because of its low computation requirement may also be used, providing the input variable is filtered beforehand.

b) Calculation of conductance and susceptance

Some distance relays measure the conductance G_s and the susceptance B_s instead of resistance and reactance. The corresponding relationships are

$$G_s = \frac{R_s}{R_s^2 + X_s^2} = \frac{I_1 \cos\varphi}{U_1}$$

$$B_s = \frac{X_s}{R_s^2 + X_s^2} = \frac{I_1 \sin(-\varphi)}{U_1}$$

(12.48)

The procedure for digitally calculating G_s and B_s is similar to that for R_s and X_s. The real and apparent components of the current are therefore derived as described in Section 12.3. In the case of the voltage, however, any of the methods described in Section 12.2 may be used.

The procedure employed to determine the conductance G_s is especially important, since the condition

$$G_s > \frac{1}{R_b}$$

(12.49)

defines the area inside the circular characteristic of diameter R_b in the R_s/X_s plane shown in Fig. 12.14. Having calculated G_s, checking the condition (12.49)

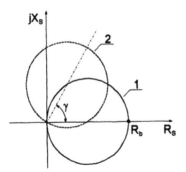

Fig. 12.14: Conductance (1) and MHO operating characteristic (2) as used in distance
relays

digitally is a simple matter, which is extremely important for the circular characteristics used in distance relays in MV power systems.

Having created the conductance circle (1) in Fig. 12.4, rotating it by the angle
γ to obtain the well-known MHO characteristic (2) does not present a problem,
e.g. the voltage signal can be delayed by r samples so that

$$r = \frac{\gamma\, m}{2\pi} \tag{12.50}$$

12.5.3 Measurement based on a line model of the first order

Assuming the simple line model of Fig. 12.15, a short-circuited transmission line
is described by the differential equation

$$u(t) = R_s\, i(t) + L_s\, i'(t) \tag{12.51}$$

While the calculation of a fault current i(t) is normally based on the values
given for R_s and X_s and the voltage curve u(t), the procedure when measuring
distance is the reverse, i.e. the voltage u(t) and current i(t) together with its derivative i'(t) are measured and the values of the unknowns R_s and X_s calculated
from them. Two independent equations are needed to do this. The different alter-

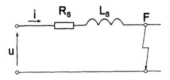

Fig. 12.15: Simple model of a transmission line

natives of the algorithms proposed up to the present depend on the methods used to formulate these two equations, some of which are described below.

a) Determination of the current and voltage values at two different times [C.11]

If equation (12.51) is solved for two different times t_1 and t_2, the following two independent equations are obtained

$$u_1 = R_s \, i_1 + L_s \, i'_1$$
$$u_2 = R_s \, i_2 + L_s \, i'_2$$

(12.52)

From which

$$R_s = \frac{u_1 \, i'_2 - u_2 \, i'_1}{i_1 \, i'_2 - i_2 \, i'_1}$$

$$L_s = \frac{u_2 \, i_1 - u_1 \, i_2}{i_1 \, i'_2 - i_2 \, i'_1}$$

(12.53)

Basically, any times can be chosen for t_1 and t_2, but to minimize computing time, the interval Δt between t_1 and t_2 should not be too large. For this reason, $\Delta T = T_i$ is frequently chosen.

The values of current, voltage and the current derivative at time t_2 are

$$i_2 = \frac{i(n) + i(n-1)}{2\cos\frac{\omega_1 T_i}{2}}$$

$$u_2 = \frac{u(n) + u(n-1)}{2\cos\frac{\omega_1 T_i}{2}}$$

$$i'_2 = \frac{i(n) - i(n-1)}{2\sin\frac{\omega_1 T_i}{2}}$$

The calculation of the corresponding values at time t_1 is the same with the exception that the expression $(n - r)$ is inserted in the above equations in place of n whereby $t_2 - t_1 = rT_i$.

Since only three consecutive samples are needed to calculate R_s and L_s, the above procedure is very fast. One of the consequences of the line model assumed is that any DC components in the current and/or voltage have no influence on accuracy. This does not apply, however, to HF oscillations which can cause significant errors and therefore low-pass filters should be inserted in the inputs of both current and voltage functions.

One of the negative aspects of this procedure is its high computation requirement which results from the large number of multiplications necessary.

b) Methods of integration

In order to reduce the adverse influence on accuracy of the HF content in the input variables, two independent equations can be derived by double integration from equation (12.51) [C.23].

$$\int_{t_1}^{t_2} u(\tau)\,d\tau = R_s \int_{t_1}^{t_2} i(\tau)\,d\tau + L_s\left[i(t_2) - i(t_1)\right]$$

(12.54)

$$\int_{t_3}^{t_4} u(\tau)\,d\tau = R_s \int_{t_3}^{t_4} i(\tau)\,d\tau + L_s\left[i(t_4) - i(t_3)\right]$$

The unknowns R_s and L_s can be determined using these equations in a similar manner to equation (12.52). The integrals are calculated either digitally or according to Euler, or by application of the trapezium rule. The influence of harmonics and HF oscillations can be minimized by appropriate choice of the limits for integration. Note that the times t_1 to t_4 must coincide with samples. As before, the line model assumed excludes any influence on accuracy of DC components contained in the input variables.

c) Squared error principle [C.2, C.22]

The differential equation (12.51) for the first order line model is modified slightly by writing

$$u(t) = R_s\, i(t) + L_s\, i'(t) - e(t)$$

(12.55)

where $e(t)$ is the error arising from the simplified representation of the transmission line (omission of the transverse capacitances and leakage resistances).

Expressing equation (12.55) in terms of the square of the error $e^2(t)$

$$e^2(t) = R_s^2\, i^2(t) + L_s^2\, i'^2(t) + u^2(t) +$$
$$+ 2\left[R_s L_s\, i(t)\, i'(t) - R_s\, i(t)\, u(t) - L_s\, i'(t)\, u(t)\right]$$

(12.56)

The optimum values for R_s and L_s coincide with the minimum value of the integral of the square of the error, i.e.

$$0.5\frac{\partial}{\partial R_s}\int_{t_0}^{t}e^2(\tau)d\tau = R_s\int_{t_0}^{t}i^2(\tau)d\tau + L_s\int_{t_0}^{t}i(\tau)i'(\tau)d\tau -$$

$$-\int_{t_0}^{t}i(\tau)u(\tau)d\tau = 0$$

(12.57)

$$0.5\frac{\partial}{\partial L_s}\int_{t_0}^{t}e^2(\tau)d\tau = R_s\int_{t_0}^{t}i(\tau)i'(\tau)d\tau + L_s\int_{t_0}^{t}i'^2(\tau)d\tau -$$

$$-\int_{t_0}^{t}i'(\tau)u(\tau)d\tau = 0$$

This method is one of the most accurate, but has a considerable computation requirement mainly because of the many floating comma multiplications.

d) Frequency domain solution

The following expression is obtained by multiplying both sides of the differential equation (12.51) by $\exp(-j\omega t)$ and then integrating it between the limits t_0 and t.

$$\int_{t_0}^{t}u(\tau)\exp(-j\omega\tau)\,d\tau = R_s\int_{t_0}^{t}i(\tau)\exp(-j\omega\tau)\,d\tau +$$

(12.58)

$$+ j\omega L_s\int_{t_0}^{t}i(\tau)\exp(-j\omega\tau)\,d\tau - L_s\,i(t)\exp(-j\omega t) + L_s\,i(t_0)\exp(-j\omega t_0)$$

The solution of this equation is based on the determination of the finite Fourier transformations calculated between the integration limits t_0 and t. The real and apparent components can be derived from equation (12.58) such that

$$\int_{t_0}^{t} u(t)\cos\omega\tau \, d\tau = R_s \int_{t_0}^{t} i(\tau)\cos\omega\tau \, d\tau +$$

$$+ L_s \left[\omega \int_{t_0}^{t} i(\tau)\sin\omega\tau \, d\tau - i(t)\cos\omega t + i(t_0)\cos\omega t_0 \right]$$

$$\int_{t_0}^{t} u(\tau)\sin\omega\tau \, d\tau = R_s \int_{t_0}^{t} i(\tau)\sin\omega\tau \, d\tau +$$

$$+ L_s \left[-\omega \int_{t_0}^{t} i(\tau)\cos\omega\tau \, d\tau - i(t)\sin\omega t + i(t_0)\sin\omega t_0 \right]$$

(12.59)

Thus two independent equations with two unknowns R_S and L_S have resulted which can be solved in the same way as the equations (12.52). Although the calculation of the factors in equation (12.59) would appear to be laborious, it is in reality uncomplicated, because every integral is a correlation of either current or voltage with sine and cosine functions. The number of absolutely essential multiplications can be reduced by applying the Euler method of integration or the infinite impulse response relationship (see Section 11.2).

Theoretically, the equations (12.59) permit every value of R_S and L_S to be determined for every value of ω and any desired data window $(t - t_0)$. In practice, however, the angular velocity ω is assumed to be approximately equal to that of the fundamental ω_1 and the data window is never less than $T_1/4$, i.e. never less than a quarter of the fundamental.

This method is fast and accurate, but has a significant computation requirement in spite of the application of the simplest possible digital equations.

12.5.4 Minimizing the influence of fault resistance on the distance measurement

Generally, there is always some resistance at the location of a fault on an overhead line. For phase faults, this is the resistance of the arc which is seldom higher than 0.5 Ω. Ground faults involve ground resistance which in the case of a fault to the pylon is the resistance of the ground conductor. If the pylons are connected by a ground conductor, the fault resistance is the equivalent resistance of all the footing resistances. Taking all these factors into account, the fault resistance will normally be in the range 1 to 10 Ω. Higher values up to a few hundred ohms are to be expected for ground faults via wooden masts or trees (lines without ground conductor), or broken conductors lying on the ground.

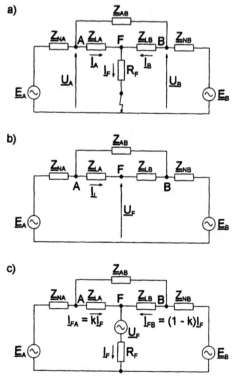

Fig. 12.16: Equivalent circuit with fault resistance at the fault location a) and equivalent circuits according to Thevenin for the normal load condition b) and
fault condition c)

The single line equivalent circuit for a three-phase fault with fault resistance
R_F at the fault location is shown in Fig. 12.16a. Figures 12.16b and 12.16c show
the equivalent circuits according to Thevenin for the normal load condition and
fault condition respectively.

Assuming the distance relay is at A and measures the current $\underline{I}_A$ and the voltage $\underline{U}_A$, the fault resistance causes the relay to measure an impedance other than
the effective one to the fault location Z_{LA}. The distance measurement thus includes an error which because of the phase-shift between the currents $\underline{I}_A$ and $\underline{I}_B$
falsifies both the resistance and reactance of the fault loop. In order to improve
the accuracy of the measurement, either a correction factor has to be introduced
or a measuring principle employed which is uninfluenced by fault resistance.

a) Correction of reactance measurement [C.28]

The impedance measured by the distance relay is

$$\underline{Z}_A = \frac{\underline{U}_A}{\underline{I}_A} = \underline{Z}_{LA} + \left(\frac{\underline{I}_F}{\underline{I}_A}\right) R_F \tag{12.60}$$

From the equivalent circuits of Fig. 12.16b and c, it follows that

$$\underline{I}_A = \underline{k}_V \underline{I}_F + \underline{I}_L$$

where k_V is the current division factor. Thus on the basis of the circuit of Fig. 12.16c

$$\underline{Z}_A = \underline{Z}_{LA} + \frac{\frac{R_F}{k_V}(\underline{I}_A - \underline{I}_L)}{\underline{I}_A \exp(j\gamma)} \tag{12.61}$$

where γ is the argument of the factor k_V which is generally close to zero.

Equation (12.61) enables both real and apparent components to be obtained separately. There are only two unknowns, the ratio R_F/k_V and the reactance of the line up to the fault X_{LA}. The latter can be derived from the resistance R_{LA} and the phase-angle of the line φ_L (see Fig. 12.17). Thus

$$X_{LA} = X_A - \frac{R_A \tan\varphi_L - X_A}{\dfrac{h_1 \tan\varphi_L}{h_2} - 1} \tag{12.62}$$

where

R_A, X_A	= resistance and reactance measured from the relay location A to the fault F
φ_L	= phase-angle of the line
h_1	= $\mathrm{Re}\big[(\underline{I}_A - \underline{I}_L)/\underline{I}_A \exp(j\gamma)\big] \rightarrow$ real component
h_2	= $\mathrm{Im}\big[(\underline{I}_A - \underline{I}_L)/\underline{I}_A \exp(j\gamma)\big] \rightarrow$ apparent component

Dividing h_1 by h_2

$$\frac{h_1}{h_2} = \frac{\mathrm{Re}\big[(\underline{I}_A - \underline{I}_L)/\underline{I}_A^* \exp(j\gamma)\big]}{\mathrm{Im}\big[(\underline{I}_A - \underline{I}_L)/\underline{I}_A^* \exp(j\gamma)\big]} \tag{12.63}$$

where $\underline{I}_A{}^*$ is the vector conjugated to the vector $\underline{I}_A$

It is therefore possible to determine the effective reactance up to the fault from equation (12.62), providing the solution of equation (12.63) is known and this is not difficult, because similar algorithms are used to digitally calculate the

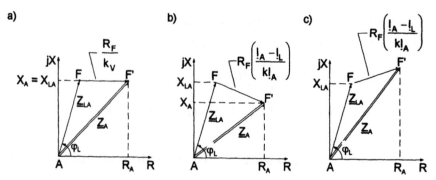

Fig. 12.17:   Influence of R_F on the distance measurement
a) fault supplied only from station A
b) fault supplied from both stations, power flows from A to B
c) fault supplied from both stations, power flows from B to A

real and apparent components as are used to determine the real and apparent powers (see Section 12.4). Note that the load current i_L no longer flows at the time of measurement and this has to be taken into account when calculating the difference $i_A - i_L$. For example, if the quotient h_1/h_2 is determined by signal correlation (see Section 12.4.1) and assuming $\gamma = 0$, the following equation results

$$\left(\frac{h_1}{h_2}\right)_n = \frac{\displaystyle\sum_{k=0}^{\frac{m}{2}-1}\left\{\left[i_A(n-k)-i_A(n-rm-k)\right]i_A(n-k)\right\}}{\displaystyle-\sum_{k=0}^{\frac{m}{2}-1}\left\{\left[i_A(n-k)-i_A(n-rm-k)\right]i_A\left(n-\frac{m}{4}-k\right)\right\}} \qquad (12.64)$$

where r is a whole number in periods of the fundamental denoting how far back reference has to be made to be absolutely sure of the load current prior to the fault, i.e. $i_L(n) = i_A(n - rm)$.

Even though equation (12.62) appreciably improves the accuracy of the distance measurement, the use of this algorithm does not meet with unanimous approval, because selectivity can also be reliably achieved by optimum choice of operating characteristic. On the other hand, means of eliminating the adverse influence of fault resistance on measurement accuracy are extremely important for fault locators based on the impedance principle.

b) Direct measurement of the distance to the fault [C.26]

Equation (12.61) can also be written in the following form

$$\frac{U_A}{(I_A - I_L)\exp(-j\gamma)} - \frac{I_A Z_{LA}}{(I_A - I_L)\exp(-j\gamma)} = \frac{R_F}{k_v} \tag{12.65}$$

Since the right-hand side of this equation is a wholly real expression, the apparent parts of the left-hand side can be written

$$Im\left[\frac{U_A \exp(j\gamma)}{I_A - I_L}\right] = Im\left[\frac{I_A Z_{LA} \exp(j\gamma)}{I_A - I_L}\right] \tag{12.66}$$

By multiplying both sides of the last equation by $(I_A{}^* - I_L{}^*)$, the impedance $Z_{LA} = I Z_1$ to the fault location is obtained where I is the distance from the relay to the fault and Z_1 is the positive-sequence line impedance per km. From equation (12.66),

$$I = \frac{Im\left[U_A\left(I_A^* - I_L^*\right)\exp(j\gamma)\right]}{Im\left[I_A Z_{LA}\left(I_A^* - I_L^*\right)\exp(j\gamma)\right]} \tag{12.67}$$

The apparent expressions in numerator and denominator can be determined using the algorithms described in Section 12.4.

c) Taking variations of the phase-angle γ into account [C.5]

The above procedures rely on the assumption that the phase-angle γ is constant; it is frequently assumed to be zero. In reality, however, the phase-angle changes in proportion to the impedance between the relay and the fault. The calculation takes this into account by varying the division of current factor k_v in equation (12.61) according to power system configuration. If, for example,

$$Z_{AB} = Z_{LA} + Z_{LB}$$

in Fig. 12.16, i.e. the fault is on one of the two lines, the following is true

$$k_v = \frac{(I - I^*)(Z_{NA} + Z_{NB} + Z_{AB}) + Z_{NB}}{2(Z_{NA} + Z_{NB}) + Z_{AB}} \tag{12.68}$$

where $I^* = Z_{LA}/Z_{AB}$, the relative distance to the fault

By inserting equation (12.68) into equation (12.61) and assuming $Z_{LA} = I^* Z_{AB}$

$$I^{*2} - I^* A + B_1 - R_F B_2 = 0 \tag{12.69}$$

where

$$\underline{A} = 1 + \frac{Z_A}{Z_{AB}} + \frac{Z_{NB}}{Z_{NA} + Z_{NB} + Z_{AB}}$$

$$\underline{B}_1 = \frac{Z_A}{Z_{AB}}\left(1 + \frac{Z_{NB}}{Z_{NA} + Z_{NB} + Z_{AB}}\right)$$

$$\underline{B}_2 = \left(\frac{I_A - I_L}{I_A}\right)\frac{2(Z_{NA} + Z_{NB}) + Z_{AB}}{Z_{AB}(Z_{NA} + Z_{NB} + Z_{AB})}$$

Since only Z_A and $(I_A - I_L)/I_A$ depend on the location of the fault, most of the calculations to determine the factors $\underline{A}$, $\underline{B}_1$ and $\underline{B}_2$ can be carried out before the fault actually occurs.

Equation (12.69) is a complex expression with the unknowns I^* and R_F. If it is split into its real and apparent parts, the following equation can be derived from the apparent part

$$I^{*2}\,\mathrm{Im}[\underline{B}_2^{-1}] - I^*\,\mathrm{Im}[\underline{A}\underline{B}_2^{-1}] + \mathrm{Im}[\underline{B}_1\underline{B}_2^{-1}] = 0 \tag{12.70}$$

the solution of which is the desired distance to the fault.

d) Utilization of the instantaneous values of the first order differential equation [C.31]

The differential equation of the line model shown in Fig. 12.16a is

$$u_A = i_A R_{LA} + i'_A L_{LA} + i_F R_F \tag{12.71}$$

Assuming that the current division factor only comprises a real component (i.e. $\underline{k}_V = k_V$), then

$$i_F = \frac{i_A - i_L}{k_V}$$

where i_L is the instantaneous load current prior to the fault.

Equation (12.71) can only be written in the following form

$$u_A = L_{LA}(i_A \cot \varphi_L + i'_A) + \frac{R_F}{k_V}(i_A - i_L) \tag{12.72}$$

where $\cot \varphi_L = \dfrac{R_{LA}}{L_{LA}\omega}$

Equation (12.72) contains two unknowns, L_{LA} and R_F/k_v. The value of L_{LA} can be determined according to the procedure described in Section 12.5.3. If the equation is solved for two different times, the following are obtained

$$u_{A1} = L_{LA}\left(i_{A1}\cot\varphi_L + i'_{A1}\right) + \frac{R_F}{k_v}\left(i_{A1} - i_{L1}\right)$$

$$\tag{12.73}$$

$$u_{A2} = L_{LA}\left(i_{A2}\cot\varphi_L + i'_{A2}\right) + \frac{R_F}{k_v}\left(i_{A2} - i_{L2}\right)$$

where

u_{A1}, i_{A1}, i'_{A1} = voltage, current and derived current at time t_1
u_{A2}, i_{A2}, i'_{A2} = voltage, current and derived current at time t_2
i_{L1}, i_{L2} = hypothetical load currents at time t_1 and t_2 respectively, had no fault occurred

Equations (12.73) only enables L_{LA} to be found as follows

$$L_{LA} = \frac{u_{A1}\left(i_{A2} - i_{L2}\right) - u_{A2}\left(i_{A1} - i_{L1}\right)}{\left(i_{A1}\cot\varphi_L + i'_{A1}\right)\left(i_{A2} - i_{L2}\right) - \left(i_{A2}\cot\varphi_L + i'_{A2}\right)\left(i_{A1} - i_{L1}\right)} \tag{12.74}$$

All the procedures discussed up to the present for reducing the influence of fault resistance on distance measurement applied to the single-line equivalent circuit for a three-phase fault. Similar equations can be derived for ground faults with the exception that the corresponding current has to be inserted for i_A depending on whether zero-sequence compensation or phase current compensation is in use. In the case of zero-sequence compensation, the balancing equation is

$$u_A = R_{LA}i_C + L_{LA}i'_D + \frac{R_F}{k_v}\left(i_A - i_L\right) \tag{12.75}$$

where

i_A = current of the line with the fault
i_L = pre-fault load current
i_C = current of the faulted conductor plus zero-sequence current i_0 to compensate zero-sequence resistance, i.e. $i_C = i_A + k_R i_0$ where $k_R = \left(R_{L0} - R_{L1}\right)/R_{L0}$
i_D = current of the faulted conductor plus zero-sequence current i_0 to compensate zero-sequence reactance, i.e. $i_D = i_A + k_R i_0$ where $k_L = \left(L_{L0} - L_{L1}\right)/L_{L0}$
R_{L0}, L_{L0} = zero-sequence resistance and inductance
R_{L1}, L_{L1} = positive-sequence resistance and inductance

Following the same procedure as for equations (12.71) and (12.72), the value of L_{LA} can be calculated from the two independent equations with two unknowns.

12.6 Digital frequency measurement

The measurement of frequency in electric power systems is mainly concerned with the fundamental and a narrow band around it. In addition to measuring the absolute value, the rate of frequency change df/dt is also used in load-shedding applications as an indication of a deficit of generating capacity and impending underfrequency.

Of the many principles for digitally measuring frequency, the most important ones are described below.

12.6.1 Frequency measurement by counting reference impulses

Counting the impulses of a reference clock during a half or a full period of the frequency to be measured is one of the obvious and oldest methods applied to measure frequency.

The following relationship applies for counting the impulses during a half-cycle of the fundamental at $f_N = 50$ Hz and 60 Hz respectively

$$f = \frac{50m}{2p} = \frac{25m}{p} \quad [\text{Hz}], \quad f = \frac{60m}{2p} = \frac{30m}{p} \quad [\text{Hz}] \tag{12.76}$$

where
m $= T_1/T_i =$ number of impulses equivalent to a period of the fundamental at rated frequency
p $=$ number of impulses counted during half a period of the fundamental

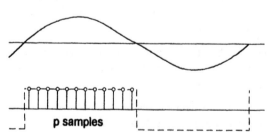

Fig. 12.18: Frequency measurement by counting the impulses of a reference clock during a half-cycle of the frequency to be measured

While this principle is very simple, the frequency of the reference clock should be high to minimize the error Δf which is given by

$$\Delta f^* = \pm \frac{2}{m}$$

Since the measuring range is small and small deviations of frequency have to be detected, the permissible error is also small.

In addition to an adequately high clock frequency, it is essential to be able to determine the zero-crossings of the voltage of the frequency to be measured accurately. This can be achieved by approximating the voltage curve between two consecutive samples of opposite polarity by a straight line. As can be seen from Fig. 12.10,

$$t_2 = \frac{T_i u(n)}{u(n) - u(n-1)}$$

$$t_1 = T_i - t_2$$

Assuming that the frequency is determined by measuring the duration of a period of the voltage waveform between two consecutive positive-going zero-crossings, the following conditions can be formulated

- If $u(n) < 0$ or $[u(n) + u(n-1)] > [u(n) - u(n-1)]$, i.e. the polarity of the voltage curve does not change between samples (n) and $(n-1)$ or only as a result of modulation, then $T_P = T_p + 1$.
- If $u(n) > 0$ or $[u(n) + u(n-1)] \le [u(n) - u(n-1)]$, i.e. the positive-going voltage curve crosses the time axis between samples (n) and $(n-1)$, then

$$g = 0.5\left[1 - \frac{u(n) + u(n-1)}{u(n) - u(n-1)}\right]$$

$$T_p = T_p + g$$

$$f = \frac{50m}{T_p} \text{ [Hz]} \quad \text{respectively} \quad \frac{60m}{T_p} \text{ [Hz]}$$

$$T_p = 1 - g$$

At a clock frequency of $f_i = 800$ Hz, the error in the region of the rated frequency does not exceed 0.01 Hz.

The rate of frequency change is given by the following relationship

$$\frac{\Delta f}{\Delta t} \cong \frac{f - f_r}{T_1}$$

where
f   = frequency of last period measured
f_r = frequency measured in the preceding period

12.6.2 Frequency measurement by convolution of a Walsh function

The deviation from the rated frequency f_N can be determined digitally by convolution of a zero order Walsh function with a data window $T_w = T_i$. For an input variable

$$u(t) = U_1 \sin(\omega_k t + \alpha) \tag{12.78}$$

where ω_k is the unknown angular velocity, the convolution of the Walsh function is of the form

$$F(t) = \frac{2U_1}{\omega_k} \sin\left(\frac{\omega_k T_1}{2}\right) \sin\left(\omega_k t - \frac{\omega_k T_1}{2} + \alpha\right) \tag{12.79}$$

Taking account of the fact that $\omega_k = (\omega_1 + \omega_d)$, in which ω_d is the additional angular velocity, equation (12.79) can be written

$$F(t) = \frac{2U_1}{\omega_k} \sin\left(\frac{\omega_d T_1}{2}\right) \sin\left(\omega_k t - \frac{\omega_d T_1}{2} + \alpha\right) \tag{12.80}$$

A sinusoidal signal is thus obtained having the same frequency as the input variable u(t) but an amplitude

$$F_1 = \frac{2U_1}{\omega_k} \sin\left(\frac{\omega_d T_1}{2}\right)$$

The following is an approximation for small deviations of frequency, i.e. in the range $\omega_d = \pm 0.1\, \omega_1$,

$$\sin\left(\frac{\omega_d T_1}{2}\right) \cong \frac{\omega_d T_1}{2}$$

for which the deviation f_d from the frequency f_1 is

$$f_d \cong \frac{F_1}{U_1 T_k T_1} \tag{12.81}$$

where
T_k = period at the angular velocity ω_k.

The curve of F(t) is roughly in phase with the input variable u(t) for a frequency increase, i.e. $\omega_k > \omega_1$, and roughly in anti-phase for a frequency decrease, i.e. $\omega_k < \omega_1$. This can be used to determine when the frequency deviation has a positive and when a negative sign.

Since the value T_k of the input variable is unknown, it is usually assumed that $T_k = T_1$ which can cause an error. However, by differentiating equation (12.78)

$$H(t) = F'(t) = 2U_1 \sin\left(\frac{\omega_d T_1}{2}\right) \cos\left(\omega_k t + \alpha - \frac{\omega_d T_1}{2}\right) \tag{12.82}$$

By applying the same simplifications as for the derivation of equation (12.18), the amplitude of the signal $H(t)$ becomes

$$H_1 \cong 2\pi U_1 T_1 f_d \tag{12.83}$$

and enables the frequency deviation f_d to be determined.

Digital integration of the signal $u(t)$ can be achieved using the trapezium rule which gives

$$F(n) = \left\{0.5\left[u(n) + u(n-m)\right] + \sum_{k=1}^{m-1} u(n-k)\right\} T_i \tag{12.84}$$

Differentiating this signal produces

$$H(n) = \frac{F(n) - F(n-1)}{T_i} \tag{12.85}$$

$$= 0.5\left[u(n) + u(n-1) - u(n-m) - (n-m-1)\right]$$

It follows from equation (12.85) that H_1 can be easily calculated, but whether the frequency deviation is positive or negative has to be determined separately, i.e. whether the frequency is above or below the rated frequency. There are several simple ways of doing this.

The above procedure has many advantages such as

- fairly accurate measurement of frequency deviation at a relatively low sampling rate
- low computing time requirement
- uninvolved computations

Its most important disadvantages are its sensitivity to distorted input variables and dependence on the accuracy of the amplitude U_1.

12.6.3 Frequency measurement by correlation with sine and cosine functions

If the frequency of an input variable given by equation (12.78) has to be measured and $T_w = T_1$, the signal can be correlated with sine and cosine functions so that

$$C_1 = \int_{t-T_1}^{t} u(\tau) \sin \omega_1 \tau \, d\tau$$

$$C_2 = \int_{t-T_1}^{t} u(\tau) \cos \omega_1 \tau \, d\tau$$

(12.86)

After inserting the voltage signal according to equation (12.78) and integrating

$$C_1 = \frac{U_1 T_1}{2} \left\{ \begin{array}{l} K_1 \cos(\omega_d t + \alpha - \beta_1) - \\ K_2 \cos[(2\omega_1 + \omega_d)t + \alpha_1 - \beta_2] \end{array} \right\}$$

$$C_2 = \frac{U_1 T_1}{2} \left\{ \begin{array}{l} K_1 \sin(\omega_d t + \alpha - \beta_1) - \\ K_2 \sin[(2\omega_1 + \omega_d)t + \alpha_1 - \beta_2] \end{array} \right\}$$

(12.87)

where

$$\omega_d = \omega_k - \omega_1$$

$$K_1 = \sin \frac{\omega_d T_1}{2} \Big/ \frac{\omega_d T_1}{2}$$

$$K_2 = \sin \frac{(2\omega_1 + \omega_d)T_1}{2} \Big/ \frac{(2\omega_1 + \omega_d)T_1}{2}$$

$$\beta_1 = \frac{\omega_d T_1}{2}$$

$$\beta_2 = \frac{(2\omega_1 + \omega_d)T_1}{2}$$

Thus every correlation coefficient is the sum of two terms, the first being the angular velocity ω_d and the second almost double the angular velocity. From Fig. 12.19 it is clear that even for a severe deviation of frequency, the influence of the component at double the angular velocity ω_1 is only slight (curve 2). This com-

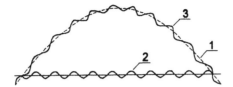

Fig. 12.19: Curves of the functions C_1 and C_2 for $\omega_d = 0.1\omega_1$
  1 = first term of the sum in equation (12.87)
  2 = second term of the sum in equation (12.87)
  3 = sum of curves 1 and 2

ponent can be completely removed by inserting a digital low-pass filter, in which case

$$C_1{}^* = 0.5KU_1T_1\cos\left(\omega_d t + \alpha - \frac{\omega_d T_1}{2} - \varphi_p\right)$$

$$C_2{}^* = 0.5KU_1T_1\sin\left(\omega_d t + \alpha - \frac{\omega_d T_1}{2} - \varphi_p\right)$$

(12.88)

where
K   = factor to compensate for the change in amplitude due to convolution
φ_p = phase-shift due to convolution (see Section 11.3.1)

By differentiating the second of these equations and dividing by the first,

$$\frac{dC_2^{\cdot}}{dt}\frac{1}{C_1^{\cdot}} = \omega_d = 2\pi f_d$$

(12.89)

Equation (12.89) is used for the case that $C_1^{\cdot} > C_2^{\cdot}$. Where this condition is reversed, greater accuracy is obtained from

$$-\frac{dC_1^{\cdot}}{dt}\frac{1}{C_2^{\cdot}} = \omega_d = 2\pi f_d$$

(12.90)

The digital forms of these relationships are

$$f_d = \frac{C_2^{\cdot}(n) - C_2^{\cdot}(n-1)}{\pi T_i\left[C_1^{\cdot}(n) + C_1^{\cdot}(n-1)\right]}$$

(12.91)

$$f_d = -\frac{C_1^{\cdot}(n) - C_1^{\cdot}(n-1)}{\pi T_i\left[C_2^{\cdot}(n) + C_2^{\cdot}(n-1)\right]}$$

(12.92)

This procedure provides high accuracy even at low sampling rates and a short computation time. It use is recommended especially in cases where the coefficients C_1 and C_2 have to be calculated in any event, e.g. when determining the amplitude of a voltage signal.

13

Logical Structures for Digital Protection

13.1 Logical function areas

One of the most striking advantages of programmable digital protection devices is the ease with which logical conditions and relationships can be checked. As a consequence of the large volume of data and logical criteria which modern digital devices can process, it has been possible to expand the information exchanged between the protected unit and the protection devices and thus increase the speed and accuracy of fault detection and location and has also enabled the protection to take on new tasks.

The logical structure of complex protection devices and systems can be divided into the following functional areas

- initial fault detection
- determination of whether the fault is in the assigned zone of protection
- adjustment of the zone of protection to suit the state of the system, e.g. reduction of the first zone of a distance relay after the first auto-reclosure attempt
- selection of the faulted phase or phases
- exchange of data with other protection devices or systems, e.g. with a protection relay at the opposite end of the line
- tripping control and communication with operating personnel
- automatic self-supervision

The following Sections are only concerned with the logical structures connected with the occurrence of a fault and its selective detection.

13.2 Logical structure of simple protection devices

This kind of structure is explained in relation to an overcurrent protection relay. The flow chart of a definite time overcurrent relay is given in Fig. 13.1. The current signal i(n) is first filtered in (1) before passing to the arithmetic unit (2) in which an appropriate algorithm derives a variable proportional to the amplitude of the current. The condition for operation is checked in (3), i.e. whether the variable proportional to the amplitude exceeds the setting A. Providing this condition is fulfilled, unit (4) supervises optional conditions, e.g. whether three consecutive samples of the signal I(n) fulfil the condition set by (3). The protection device trips (unit (5)) if the optional condition is also fulfilled. As soon as the condition of (3) is not fulfilled, unit (6) checks the status of (5). If (6) determines that the relay has operated, the condition for not resetting is checked in (7) in relation to the setting B. Providing the condition is fulfilled the tripped status is maintained; if it is not, (7) resets the relay.

Figure 13.2 shows the corresponding flow chart of a time-overcurrent device. As before, the pick-up condition is determined in (3) in relation to the setting A. The timing unit (4) is excited as soon as the signal proportional to the input variable exceeds A. The delay of (4) depends on the time/current characteristic chosen for the particular device. For example, a hyperbolic characteristic is obtained

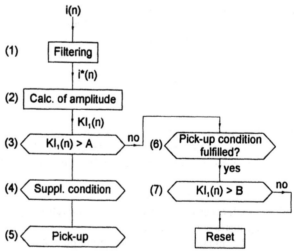

Fig. 13.1:   Flow chart of a digital instantaneous overcurrent device

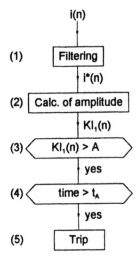

Fig. 13.2:   Flow chart of a digital time-overcurrent device

by setting the condition

$$\sum_{k=r}^{n}\left[KI(k) - A\right] \geq C$$

where
C   = term defining the characteristic
r   = sample fulfilling the pick-up condition of (3)
The relay trips as soon as the time condition of (4) is fulfilled.

The examples of logical structures discussed above may vary from one protection device to another, but the basic principle is always the same.

13.3 Logical structure for determining the operating characteristic of a distance relay

The logical structure of a distance protection device is especially complicated, because it can include alternative configurations in addition to the functional areas listed at the beginning of the Chapter. Only one functional area, the one directly concerning the fulfilment or non-fulfilment of the condition for operation of a digital distance relay, is described below. This involves determining whether the measured resistance and reactance of the protected line lie inside or outside the operating characteristic shown in Fig. 13.3a. The diagram assumes a two

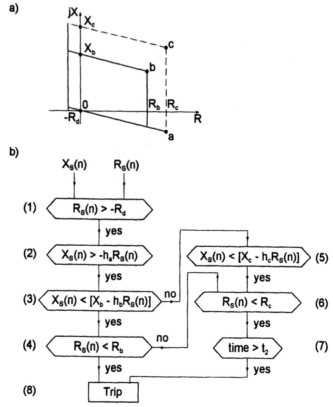

Fig. 13.3: Operating characteristic a) and flow chart b) for checking the fulfilment of the condition for operation of a distance protection device

zone relay with the first zone undelayed ($t_1 = 0$) and the second zone delayed by the time t_2.

One possible logical structure of a distance protection device is shown in Fig. 13.3b. Unit (1) supervises whether the measured resistance R_s is to the right of the straight line $R = -R_d$. Similarly, (2) determines whether the measured reactance X_s is above the straight line (O, a) with the directional factor $-h_a$. Providing both these conditions are fulfilled, (3) checks that both resistance and reactance are below the first zone limit line (X_b, b) with the directional factor $-h_b$. If this condition is also fulfilled, (4) checks whether the resistance R_s is less than the first zone resistance limit line R_b. If this last condition is also fulfilled, then the fault must lie in the first zone and the distance relay trips. Assuming the condition determined by (3) is not fulfilled, (5) checks whether the reactance is below the limit line (X_c, c) for the second zone. Should this condition be fulfilled,

(6) determines whether the resistance R_S is to the left of the second zone limit line R_c. As soon as all the conditions for the second zone are fulfilled, the time t_2 in (7) starts to run and tripping takes place in (8) when it expires.

As pointed out, this is only one of several possible structures which can be expanded to measure more zones or to include other conditions. To minimize the computation requirement, the directional factors h_a, h_b and h_c should be chosen to keep the number of multiplication operations to a minimum.

13.4 Logical structure for transformer differential protection devices

The logical structure of a differential protection device is a typical example of how the sensitivity and operating speed of a protection relay can be improved by the simpler access to information (e.g. voltage values and tap-changer position) and the ease of processing logical data, both inherent characteristics of digital computer systems.

The flow chart of a differential relay for protecting power transformers can be seen from Fig. 13.4. The chart has been simplified by showing only one phase. The input variables and information comprise the

- currents $i_a(n)$ and $i_b(n)$ measured on both sides of the transformer
- phase or phase-to-phase voltages $u_a(n)$ and $u_b(n)$ measured on both sides of the transformer
- signal h_p indicating the position of the tap-changer

The operating and restraint current components $i_\Delta(n)$ and $i_{rstr}(n)$ are derived in (1). The result for the operating current is multiplied by h_p to take account of the tap-changer position and the effective ratio of the protected transformer. This minimizes the influence of the changing transformer ratio. (2) generates the operating characteristic as a function of the restraint current. Here, the additional information regarding the position of the tap-changer enables the characteristic to be optimized with respect to the lowest permissible basic setting and the pick-up ratio. (3) then checks that the operating current is not a false one caused by c.t. saturation during a through-fault [C.30]. Once this possibility has been eliminated, (4) determines whether the delayed operating current is the result of an increased power transformer magnetizing current. After this has also been excluded, (5) checks that it is not in reality a transformer inrush current. The latter can be excluded, if a voltage change was measured immediately beforehand. In a similar way, (6) examines the possibility of power transformer saturation due to overvoltage or underfrequency. The supplementary condition of (7) determines how many consecutive samples with agreeing results are necessary before tripping can take place. If magnetizing current cannot be excluded by (5) and (6) as

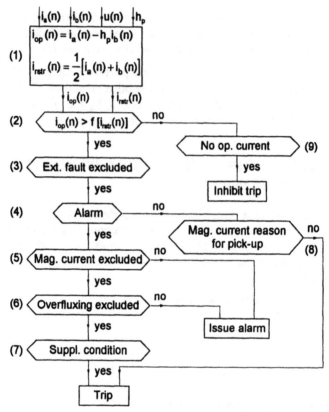

Fig. 13.4:   Flow chart of a digital differential protection device for a power transformer

the possible cause of the differential current i_Δ, all functions are blocked by (4) and (8) checks more precisely on the basis of the second or fifth harmonic content or other less familiar means, whether the operating current is indeed the result of excessive magnetizing current [C.30]. Only if the existence of a fault is unequivocally confirmed does tripping take place. The relay is returned to its initial normal load operating state should (9) detect that the operating current does not occur for a time which should not be less than one period of the fundamental.

Analysis of the structure illustrated in Fig. 13.4 discloses the following positive features of differential protection

- high sensitivity which also enables inter-turn faults to be detected
- short tripping times, because high magnetizing currents can generally not occur during phase faults
- tripping only delayed by approximately 16 ms where doubt exists which necessitates more detailed harmonic analysis

13.5 Design of digital protection and control devices

13.5.1 Hardware

The general arrangement of the hardware for integrated protection and control devices can be seen from Fig. 13.5. The analogue input variables, e.g. current and voltage, are electrically insulated from the internal circuits by input transformers with grounded screens between primaries and secondaries. The analogue anti-aliasing filter, a low-pass filter as described in Section 10.2, is formed by the sampling rate chosen for the A/D converter which determines the frequency band to be transmitted. Since the multiplexer scans the input variables sequentially, which up to this point (see Section 10.3) were and subsequently have to be processed in parallel (phase relationship of currents and voltages), the multiplexer is equipped with a "sample and hold" memory according to Fig. 10.4. The values corresponding to the input variables which are now in serial form are converted to serial digital signals 4 to 16 times per period by the A/D converter. All these processes are controlled by the front end CPU (central processing unit).

The signals now pass via a DPM (dual port memory) and a digital filter (see Section 11) to the main CPU which runs the programmed protection algorithms. Depending on the scope of the protection tasks to be performed, several main CPU's may operate in parallel on a bus. The results are then communicated to a logic processor via a further DPM. The purpose of the DPM in this case is to separate the functions of the two processors as much as possible. The logic processor derives from the results supplied by the main CPU and from digital inputs,

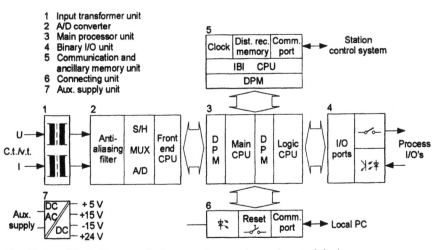

Fig. 13.5:   Block diagram of a integrated protection and control device

interlocks and settings the commands needed for tripping, signalling etc. The command signals go to the I/O (input/output) units which transmit them to the process. As with the input signals, the output signals are also electrically insulated either by output relays or opto-couplers from the process wiring.

Interfaces for the local control unit (usually an integrated key-pad with display or a PC) and station unit, i.e. the OBI (object bus interface), are also connected to the main CPU. The OBI attaches a time marker to all the data requested from the device. The values of the analogue and digital variables stored in the device before, during and after a disturbance can be accessed by either the local control unit when the device is operating in a stand-alone mode, or via the OBI when it is connected to a station control system.

The power supply unit provides the electrical insulation between the device and the station battery as well as generating all the auxiliary supplies it needs. The function of the power supply unit as well as all the other parts of the device is continuously supervised as described in Section 10.4. An example of a self-supervision system for the arrangement shown in Fig. 13.5 is given in Fig. 13.6. The following are supervised in addition to the auxiliary supplies and the input currents and voltages

- A/D converter by continuously switching reference voltages
- memories by test write/read cycles (background function)
- data communication by error codes
- processors by watchdogs

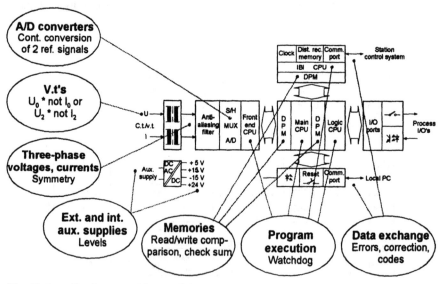

Fig. 13.6:   Continuous self-supervision

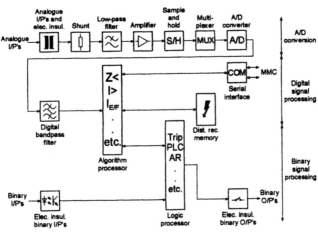

Fig. 13.7:   Flow of data and signals in an integrated protection and control device

13.5.2 Signal flow

The flow of data and signals for the example shown in Fig. 13.5 can be seen from Fig. 13.7. Input transformers electrically insulate the analogue input variables from the internal circuits before shunts convert the currents into proportional voltages. The signals pass in parallel via the anti-aliasing (low-pass) filter, amplifier and sample and hold memory to the multiplexer which converts them to serial signals. From the output of the A/D converter, all data are transferred and processed serially. The signals are processed in accordance with the protection algorithms in the main CPU which then passes the results to the logic processor where they are related to other input signals and settings before being transformed into output commands for the process. Data for evaluating disturbances and control data needed by the station control system are taken directly from the main CPU.

13.5.3 System architecture

The system architecture for the example shown in Fig. 13.5 is given in Fig. 13.8. The diagram shows the same processors and memories in greater detail and distinguishes between the various types.

    RAM (random access memory) units are used for the main memory which permit data to be written, read and changed.

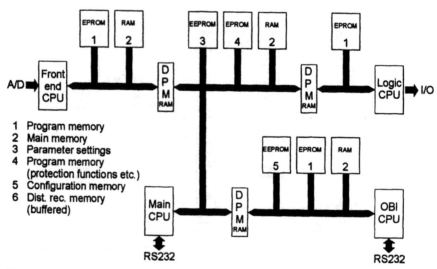

Fig. 13.8: System architecture

The application program and algorithms are contained in EPROM's (erasable programmable read-only memories). These can only be read in normal operation. The data stored in EPROM's remains intact should the auxiliary supply fail. The data is installed in the supplier's works and normally only modified there.

Settings and the information concerning protection configuration are stored in EEPROM's. This data can be read and (normally after entering a password) also changed. Data in EEPROM's also remains intact should the auxiliary supply fail.

13.5.4 Operator control

The operator controls the device either via its local key-pad with one or two-line display or by means of a PC (personal computer) connected to the interface mentioned above. Provision is also made for automatic control by a computer connected to the station control system. The latter two alternatives facilitate menu-guided dialogue with the device and more sophisticated displays than are possible with the first. Such devices are always equipped, however, with light-emitting diodes (LED's) which independently signal the most important information.

The currents and voltages measured and possibly the real and apparent powers calculated from them can be viewed in addition to the settings and the responses of the protection device. In the case of the PC and computer based methods of control, the permissible ranges are displayed together with the settings. Also the stored values of analogue and digital signals before, during and after a disturbance can be viewed. There is usually also provision to print these data where necessary. The data can also be in the form of diagrams of any desired scale and layout (e.g. superimposed zero lines) to aid evaluation.

List of settings can be produced with a time stamp assigned to every change and signal. New sets of settings can be prepared off-line on the PC or computer and then downloaded to the device either locally or via a communications channel. Resetting can also include the arrangements of the signalling and tripping logics.

13.6 Adaptive protection systems

Adaptive protection systems permit the protection functions to be adapted automatically in real time to changing power system conditions. Such procedures maintain optimum protection quality and performance at all times, but necessitates that the protection receives information from sources to which it normally does not have access [1.8].

Two conditions must be satisfied before such systems can be applied:

- The protection function (relay) has to be capable of accepting and responding to the information.

- The external information such as the positions of switchgear etc., must be accessible and in a suitable form for transfer to the protection.

Since the advent of digital relays and station control systems, both of these conditions are normally fulfilled.

Of course, all adaptations must be precisely recorded in relation to time and the external information as well as the means of its transmission must conform to the same high standards of security and reliability as those of the protection itself.

Although the principle of adaptive protection is by no means new and was applied to a limited extent with analogue relays, the current schemes with digitally operating relays and station control systems are the first instance of the functions of data acquisition, data transmission and the protection all being performed by the same techniques, at the same level of reliability, with the same self-supervision systems and in real time.

Examples of adaptive protection are:

- adjustment of the sensitivity of a biased differential function for a power transformer in relation to tap-changer position (see Section 3.2)

- changing the settings of a distance function on a double-circuit line according to the status of the parallel circuit (see Section 7.1.2.4)

- varying the load-shedding program of a frequency relay to take account of the actual distribution of real power load between the feeders to be tripped

- modifying an auto-reclosure program according to the status of the parallel circuit

- changing the settings of a distance function controlling the bus-tie breaker of a bypass busbar to suit the line it has to protect.

13.7 The application of expert systems

As digital protection devices become more and more integrated to form complex protection and control systems, there is more scope for making greater use of the data generated by the protected unit as well as its environment. The protection and control systems in a sub-station are a typical example which today exchange data in both directions. The volume of the information available and the scope and complexity of the tasks to be performed are obviously much higher at the highest and most important level, i.e. the station control level, than at the level of the individual protected units below it.

In this respect, the application of expert systems [C.44, C45, C46] for handling and processing the high volumes of protection and supervisory control data promises to be a significant step forward. Their principle is based on the procedure followed by an expert before reaching an important decision. Typical for the procedure of an expert are among other things:

1. An expert initially analyses every case in its broadest sense to take account of as many factors and boundary conditions as possible which could bear an influence on the process under examination. The decision is then taken on the basis of criteria which are dependent on the variations in the conditions.

2. The expert makes use of present data as well as data from the past and the consequences for the succeeding state.

3. The analysis must take account of the trends in data changes and their mutual interdependencies. Only the area is examined which influences the decision to be taken.

4. When an expert verifies a hypothesis he examines criteria of different weightings with respect to the decision.

5. In the event that the criteria produce contradictory or ambiguous results, the expert bases his decision on the conditions at a point in time at which the probability of an incorrect decision is reduced to a minimum.

6. The action taken to verify the hypothesis depends on the results obtained beforehand, the access to data and the urgency of the decision to be taken.

7. Where the data are incomplete, the expert arrives at his decision on the basis of his own experience and conviction in consideration of the probability of events and an estimation of wrong decisions.

Attempts to date to apply expert systems in protection have failed because of the complexity of the application software on the one hand, and the strict requirements regarding the speed with which protection decisions have to be taken on the other . Nevertheless, the application of certain elements of expert systems will without doubt become advantageous and unavoidable in the not too distant future even where protection tripping times are shorter than 20 ms. This will also apply in cases in which the information coming from the protected unit are incomplete, for example, because of c.t. saturation.

Appendix I: Symbols

Symbol	Description	Symbol	Description
input transformer	input transformer	&	AND gate
input transformer and rectifier	input transformer and rectifier	&	AND gate with one inverted input
rectifier	rectifier	≥1	OR gate
DC/DC converter	DC/DC converter	1	exclusive OR gate
A/D converter	A/D converter	&	NAND gate
bandpass filter	bandpass filter	≥1	NOR gate
band reject filter	band reject filter	1	inverter
high-pass filter	high-pass filter	timer with fixed setting	timer with fixed setting
low-pass filter	low-pass filter	pick-up delay	pick-up delay
level detector (trigger)	level detector (trigger)	reset delay	reset delay
amplifier	amplifier	timer with variable setting	timer with variable setting
adder	adder	S R	bistable flip-flop (S = set input; R = reset input)

Appendix II:
Characteristics of standard low-pass filters

The transfer functions of typical standard low-pass filters are know from the literature. They differ in the method of approximation and the order of the polynomial in the numerator. The most familiar methods of approximation are those according to Butterworth, Tschebyscheff and Bessel (Thompson).

The Butterworth, Tschebyscheff and Bessel filters of second, third and fourth orders are listed in Table II.1. The reject frequency for all these approximations is $\omega_g = 1$ rad/s. To illustrate the difference between the three methods of approximation, their amplitudes are given in Fig. II.1 for third order filters. Taking the narrowness of the transition between acceptance and reject regions as a measure of filter quality, the Bessel approximation is superior to the other two.

The most important characteristic of filters in protection is their dynamic performance, i.e. their step response. Figure II.2 shows the step responses for the same three filters shown in Fig. II.1. As is to be expected, the Bessel filter is once again the best, i.e. it has the shortest settling time and the smallest overshoot.

In order to design a low-pass filter with a cut-off frequency other than 1 rad/s, the expression $s = s_1/\omega_{ga}$ must be inserted into the transfer function of the corresponding type of filter, where s_1 is the new value of the Laplace operator and ω_{ga} the filter's cut-off angular velocity. Thus the scale of the ω axis is multiplied by ω_{ga}. This change of frequency bears a significant influence on its step response, because the scale of the time axis is also changed.

Table II.1   Typical transfer functions of low-pass filters

Order	Approx.	Transfer function
second	Be	$$\dfrac{3}{s^2 + 3s + 3}$$
	Ts	$$\dfrac{143}{s^2 + 142s + 152}$$
	Bu	$$\dfrac{1}{s^2 + \sqrt{2}s + 1}$$
third	Be	$$\dfrac{15}{s^3 + 6s^2 + 15s + 15}$$
	Ts	$$\dfrac{0.716}{s^3 + 125s^2 + 153s + 0.716}$$
	Bu	$$\dfrac{1}{s^3 + 2s^2 + 2s + 1}$$
fourth	Be	$$\dfrac{105}{\left(s^2 + 5.79s + 9.14\right)\left(s^2 + 4.22s + 1149\right)}$$
	Ts	$$\dfrac{0.358}{\left(s^2 + 0.35s + 1.06\right)\left(s^2 + 0.84s + 0.36\right)}$$
	Bu	$$\dfrac{1}{\left(s^2 + 0.76s + 1\right)\left(s^2 + 185s + 1\right)}$$

Be = Bessel,   Ts = Tschebyscheff,   Bu = Butterworth

The scale along the time axis of Fig. II.2 is not now in seconds, but in $1/\omega_{ga}$ which means the widening the acceptance region of the filter ($\omega_{ga} > 1$) shortens the settling time of the step response.

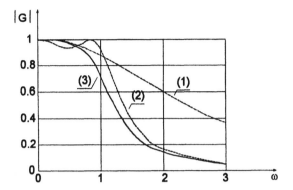

Fig. II.1:   Frequency response of third order low-pass filters with approximations according to Bessel (1), Tschebyscheff (2) and Butterworth (3)

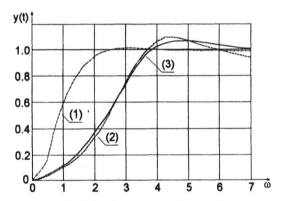

Fig. II.2:   Step response of third order low-pass filters with approximations according to Bessel (1), Tschebyscheff (2) and Butterworth (3)

References

Chapter 1

1.1   H. Ungrad: Criteria for determining the amount of protection required; Brown Boveri Review, No. 6, 1978, P. 345-347

1.2   H. Ungrad: Methods of ensuring the reliable performance of protection equipment; Brown Boveri Review, No. 6, 1978, P. 348-357

1.3   H. Hubensteiner, H. Ungrad: Reliability of protection systems; Brown Boveri Review, No. 1/2, 1983, P. 111-116

1.4   H. Ungrad, G. Schaffer, V. Narayan: A modern supervisory and protection concept for enhancing power system security; Brown Boveri Review, No. 9/10, 1984, P. 400-404

1.5   G. Böhme, F. Frey: Redundant protection for EHV networks; Brown Boveri Review, No. 3, 1980, P. 196-203

1.6   H. P. Tubandt: Selektivschutz im Netz - ein Beitrag zur Versorgungssicherheit. ETG - Fachberichte Selektivschutz; VDE Verlag GmbH, Berlin - Offenbach, 1983, P. 7-21

1.7   H. Ungrad, K. P. Brand, Cadotsch.: Interaction of machine and line protection within a coordinated substation control system; IEEE Transactions on Power Systems, 7,2 (May 92), P. 921-926

1.8   J. S. Thorp, S. H. Horowitz, A. G. Phadke: The application of an adaptive technology to power system protection and control; CIGRE 1988 Session, 34-03

Chapter 2

2.1   VDE-Vorschriften No. 0102: Leitsätze für die Berechnung von Kurzschlußströmen

2.2   A. Hochrainer: Symmetrische Komponenten in Drehstromnetzen; Springer Verlag, Berlin, Göttingen, Heidelberg, 1957

364

2.3 J. Zydanowicz: Podstawy zabezpieczen elektroenergetycznych, Volume 1, WNT Warsaw, 1979

2.4 G. Funk: Symmetrische Komponenten; Elitera Verlag, Berlin, 1976

2.5 H. Happold, D. Oeding: Elektrische Kraftwerke und Netze; Springer Verlag, Berlin, Heidelberg, New York, 1978

2.6 P. M. Anderson: Analysis of faulted power systems; The Iowa State University Press, 1973

2.7 W. Winkler, A. Przygrodzki: Zeitkonstanten von Gleichstromgliedern in Kurzschlußströmen; Zeszyty Naukowe Pol. Sl., Elektryka, No. 31, Gliwice, 1971

2.8 W. Winkler, Z. Dawid, Z. Pilch, P. Sowa: Performance of compensated transmission lines; Proc. of the 9th. PSCC, Cascais, 1987, P. 732-738

2.9 C. Urbanke: Echtzeitsimulation von Ausgleichsvorgängen in elektrischen Netzen; etz-Archiv, Volume 5, 1981, P. 141-147

2.10 G. Brauer: Elektromagnetischer Ausgleichsvorgang bei Erdschluß im erschlußkompensierten Drehstromnetz; Dissertation University of Stuttgart, 1978

2.11 H. Pundt: Untersuchung der Ausgleichsvorgänge bei Erdschluß in Energieversorgungsnetzen; Energietechnik, Volume 10, 1965, P. 469-477

2.12 A. M. Fedoseev: Relaisschutz für Elektroenergiesysteme; Energija, Moscow, 1976

2.13 H. Kugler: Schäden an Turbogeneratoren; Der Maschinenschaden, Volume 5, 1972, P. 179-188

2.14 GEC Measurements: Protective relays application guide, Stafford, 1975

2.15 B. Broniewski, W. Winkler: Die Werte von Gegenstromkomponenten in Synchrongeneratoren bei externen unsymmetrischen Fehlern; Zeszyty Naukowe Pol. Sl., Elektryka, No. 85, P. 19-30

2.16 Z. Pilch, W. Winkler: Phase comparison and distance protection performance during simultaneous double faults, IEE Conf. Publ. No. 249, London 1985, P. 42-45

2.17 J. J. Arnold: The protection of generators against negative sequence current; IEE Proc. "Developments in power system protection", London, 1975, P. 50-56

2.18 P. G. Brown: Generator It^2 requirements for system faults; IEEE Paper T 73-045-2

2.19   E. F. Knütter, K. Nimes: Elektronischer Überlastschutz 7SK2; Siemens
       Zeitschrift, Volume 8, 1976, P. 551-557

2.20   Z. Pilch, W. Winkler: Schutz- und Automatisierungskriterien für indu-
       strielle Verteilstationen mit Eigenerzeugung; II International Conference
       "Automation and Control of Distribution Systems", Tabor, 1984, P. 178-
       185

2.21   W. Leonhard: Regelung in der elektrischen Energieversorgung; Teubner
       Studienbücher, Elektrontechnik, Stuttgart, 1980

2.22   F. X. Vieira, H. V. Prado, A. G. Massaud, G. Nery: National scheme for
       controlling emergencies in the Brazilian interconnected system; CIGRE,
       Paris, 1984, Report 34-04

Chapter 3

3.1   Electricity Council: Power System Protection, Volume 1, Principles and
       Components, Peter Peregrinus Ltd., Stevenage UK and New York, 1981

3.2   H. Clemens, K. Rothe: Relaisschutztechnik in Elektroenergiesystemen;
       VEB Verlag Technik, Berlin, 1980

3.3   V. Narayan, H. Ungrad, F. Ritter: Frequency relays for load-shedding;
       Brown Boveri Review, No. 6, 1978, P. 413-415

3.4   P. G. Harrison: Considerations when planning a load-shedding program;
       Brown Boveri Review, No. 10, 1980, P. 593-598

3.5   W. Winkler, Z. Pilch: Automatic load-shedding using the rate-of-change
       of frequency criterion for large industrial power systems; Proc. of the
       19th. UPEC, Dundee, 1981, Report 17.2

Chapter 4

4.1   A. Wiszniewski: Przekladniki w elektroenergetyce; WNT Warsaw, 1982

4.2   VDE Vorschriften No. 0414: Bestimmungen für Meßwandler

4.3   M. Chamia: Transient behaviour of associated instrument transformers
       and high-speed distance and directional comparison protection; CIGRE
       Colloquium in Melbourne, 1979

4.4   W. Winkler: Some factors affecting the performance of current phase
       comparison protection; Proc. of the 20th. UPEC, Huddersfield, 1985, P.
       453-456

4.5   W. Winkler, Z. Dawid, P. Sowa: Interference harmonics in differential protection schemes caused by saturated current transformers; Proc. of the 21st. UPEC, London, 1986, P. 146-149

4.6   A. Wiszniewski, W. Winkler, P. Sowa, S. Marczonek: Selection of settings of transformer differential relays; IEE Conference Publication No. 249, London, 1985, P. 204-208

4.7   A. Van der Wal, N. Korponay: Designing current transformers for generators; ABB Relays Publication CH-ES 45-11 E

4.8   G. Rosenberger: Stromwandler mit Linearkernen; Siemens Zeitschrift 48, No. 1, 1974, P. 51-53

4.9   N. Korponay, H. Ungrad: The requirements made of current transformers by high-speed protective relays; ABB Relays Publication CH-ES 45-10.1E

4.10   W. Winkler: Sensitivity assessment of algorithms for digital fault location on interference signals in current and voltage; Prace Naukowe Inst. Energoelektryki Pol. Wrocl., No. 66, Wroclaw, 1985, P. 41-47

4.11   H. J. Freygang: Meßwandler für den Selektivschutz; ETG-Fachberichte, No. 12, VDE-Verlag GmbH, Berlin - Offenbach, 1985, P. 156-171

4.12   L. Mouton: Non-conventional current and voltage transformers, CIGRE CE/SC 34, 1980

4.13   A. Bogucki, Z. Pilch, W. Winkler, N. Korponay: Die Übertragung von hochfrequenten Schwingungen durch Spannungswandler; Bull. SEV, No. 12, 1971, P. 589-594

4.14   A. J. Schwab: Hochspannungsmeßtechnik; Springer Verlag, 1981

4.15   A. Wright: Current transformers - their transient and steady-state performance; Chapman and Hall Ltd., London, 1968

4.16   J. Ermisch: 20 Jahre Forschung auf dem Gebiet nichkonventioneller Wandler; Elektrie, Volume 43, No. 6, 1989, P. 205-207

4.17   W. Winkler: Verhalten von Vergleichsschutzsystemen bei transienten Stromwandlersättigungen; etz-Archiv, Volume 10, No. 10, 1988, P. 325-329

4.18   W. Brendler, H. Koettniz: Meßmittel und Meßmethoden in der energetischen Elektrotechnik unter dem Einfluß der Halbleiter und Mikroelektronik; Elektrie, 1984, P. 144-148

4.19   I. De Mesmaeker, Steiner: Behaviour of distance relay LZ95 in cases of extreme saturation of the main current transformers; Brown Boveri Review, No. 2, 1985, P. 78-81

Chapter 5

5.1   M. Walter: Die Entwicklung des selektiven Kurzschlußschutzes; Elektrizitätswirtschaft, Volume 11, 1967, P. 317-323

5.2   H. Neugebauer: Gleichstrom-Drehspulrelais mit Gleichrichter für die Selektivschutztechnik; ETZ, Volume 11, 1950, P. 389-393

5.3   H. Gutmann: 25 Jahre Trockengleichrichter in der Selektivschutztechnik; AEG Mitteilungen, Volume 1, 1963, P. 1-4

5.4   J. Blackburn: Protective Relaying; Marcel Dekker, New York, 1994

5.5   L. Müller: Selektivschutz elektrischer Anlagen; VWEW, 1971

5.6   VDEW-Ringbuch Schutztechnik 2: Richtlinien für statische Schutzeinrichtungen, 1983

5.7   P. Magajna: Modures, a modular protection system; Brown Boveri Review, No. 1, 1980, P. 45-49

5.8   D. Fahrenkrog-Petersen, H. Leibold: Selektivschutz in Subsystemtechnik; Siemens Energietechnik, Volume 3, P. 101-104

5.9   AEG-Telefunken: Bausteinsystem für die Schutztechnik; Part 1

5.10   Siemens Publication: Statischer Phasenvergleichsschutz 7SD31; Descriptions E14041 - F4225 - U211 - A2

5.11   Protection systems using telecommunication; CIGRE CE/SC 34/35, 1987

5.12   J. Zydanowicz, M. Namiotkiewicz: Automatyka zabezpieczeniowa w energetyce; WNT Warsaw, 1983

5.13   A. Wiszniewski, W. Winkler, , P. Sowa, S. Marczonek: Selection of settings of transformer differential relays; IEE Conference Publication No. 249, London, 1985, P. 204-208

5.14   U. Tieze, Ch. Schenk: Halbleiter-Schaltungstechnik; Springer Verlag Berlin - Heidelberg - New York - London - Paris - Tokyo, 1986

5.15   St. Dominko, J. Wroblewski: Phasenkomparatoren mit integrierten Schaltungen zum Bau von hochempfindlichen Phasenrelais; 3rd. International Scientific Conference "Aktuelle Probleme der Schutztechnik", Gliwice, 1979, Volume 1, P. 52-56

5.16 AEG-Telefunken Publication: Bausteine für die Schutztechnik, Parts 1-6.1, DC/DC-Wandlerplatten NB10, NB21, NB51

5.17 Siemens Publication: Schutztechnik, Geräte und Gerätekombinationen; Catalogue R, 1986

5.18 O. Lanz, R. Schilling, B. Kulendik: The power supply - an important link in the protection system; Brown Boveri Review, No. 2, 1981, P. 61-63

5.19 ABB Relays Publication CH-ES 66-90 E: DC/DC converters NF 92, NF 93, NF 94

5.20 AEG-Telefunken description E25.02.264.07.84. DE, Auslöseverteiler AL 400

5.21 ABB Relays Publication CH-ES 66-47 E: Matrix plug-in unit EL 91

5.22 ABB Relays Publication CH-ES 66-46 E: Opto-coupler unit Type EB 91

5.23 P. Müller, P. G. Harrison: A systematic approach to testing and supervising solid-state protection equipment; Brown Boveri Review, No. 2, 1981, P. 64-69

5.24 VDEW-Richtlinien für den Schutz von Hochspannungs-Asynchronmotoren in Kraftwerken, 1984

5.25 M. Fiorentzis: New, fully automatic means of testing generator protection equipment; Brown Boveri Review, No. 2, 1977, P. 102-107

Chapter 6

6.1 VDEW-Richtlinien für den Schutz von Hochspannungs-Asynchronmotoren in Kraftwerken, 1984

6.2 Electricity Council: Power System Protection, Volume 2, Systems and Methods, Peter Peregrinus Ltd., Stevenage UK and New York, 1981

6.3 ABB Relays Publication CH-ES 60.2 E: Products for control systems — solid-state relays and protection systems

6.4 Siemens Publication: Generator-Differentialschutz 7UD71

6.5 A. Wiszniewski, W. Winkler, P. Sowa, S. Marczonek: Selection of settings of transformer differential relays; IEE Conference Publication No. 249, London, 1985, P. 204-208

6.6 V. Narayan: Distance protection of HV and EHV transmission lines; ABB Relays Publication CH-ES 35-30.17 E

370   References

6.7   W. Kolb, I. De Mesmaeker: New distance relay for medium and high-voltage systems; Brown Boveri Review, No. 3, 1980, P. 188-195

6.8   H. Hager, C. J. Pencinger: A new approach to medium voltage transmission line protection; Western Protective Relay Conference, Spokane, Washington, 1982

6.9   ASEA Relays: Distance relay RAZOA; Application Manual AM03-7012E, Edition 3

6.10   Statischer Distanzschutz für Hochspannungsnetze — einsystemig im Subsystemgehäuse 7SL27; Siemens Gerätebuch C73000 - G1100-C46-1

6.11   GEC Measurements: Switched distance relay scheme Type SSMM3T; Publication R-5245C

6.12   AEG-Telefunken Publication: Schnelldistanzschutz SD 135A-G93 für Hochspannungsnetze

6.13   E. Zurowski: Neuer, sechsystemiger Distanzschutz für Hochspannungs. und Höchstspannungsnetze; Siemens Zeitschrift, Volume 4, 1974, P. 280-284

6.14   G. Ziegler: Mehrsystemiger Distanzschutz für Höchstspannungsnetze; Siemens Energietechnik, Volume 4, 1979, P. 153-156

6.15   W. Kolb, F. Ilar, G. Bacchini: A new distance relay for HV and EHV systems; Brown Boveri Review, No. 10, 1980, P. 599-607

6.16   AEG-Telefunken Publication: Distanzschutzeinrichtung für Hochspannnungsnetze SD 335 A

6.17   ASEA Relays: Distance relay system Type RAZFE

6.18   W. Halama: Anregeprobleme des Selektivschutzes und ihre Lösung; ETG-Fachberichte No. 12, Selektivschutz, VDE-Verlag GmbH, Berlin - Offenbach, 1983, P118-132

6.19   Siemens Publication: Dreisystemiger statischer Distanzschutz für Hochspannnungsnetze 7SL40

6.20   I. De Mesmaeker, S. Reinhard: The choice of reference voltages for the measuring systems of distance relays; Brown Boveri Review, No. 2, 1981, P. 94-101

6.21   H. Barchetti, V. Narayan: Behaviour of distance relays under close-up three-phase faults; Brown Boveri Review, No. 8/9, 1970, P. 343-347

6.22   H. Rijanto: Verhalten statischer Distanzschutzeinrichtungen mit polygonaler Auslösekennlinie bei Stromwandlersättigungen durch symmetrische Kurzschlußströme; Elektrizitätswirtschaft, Volume 21, 1979, P. 832-837

6.23   I. De Mesmaeker, A. Otto, Ch. Stein: The new distance relay LZ95 for transmission systems; Brown Boveri Review, No. 9/10, 1983, P. 379-383

Chapter 7

7.1   Protection systems using telecommunication; CIGRE CE/SC 34/35, Paris 1987

7.2   ABB Publication: Differential line protection with fibre optics Type LD91-F

7.3   AEG-Telefunken Publication: Leitungs-Differentialschutzeinrichtung SQL

7.4   Siemens Publication: Widerstandsstabilisierter Leitungsdifferentialschutz in Einschubtechnik ES902, Type 7SD12

7.5   Siemens Publication: Elektronischer Leitungsdifferentialschutz mit zwei Hilfsadern, Type 7SD20

7.6   M. Ilar: A new transformer differential relay and pilot wire relay; Brown Boveri Review, No. 2, 1981, P. 70-78

7.7   Siemens Publication: Statischer Leitungsdifferentialschutz mit zwei Hilfsadern Type 7SD74

7.8   ASEA Publication: Pilot wire differential protection Type RADHL

7.9   D. Nanko, A. Sanocki, R. Tebich, W. Winkler: A new segregated phase comparison protection scheme for EHV transmission lines; CIGRE SC 34 Meeting and Colloquium, Turku, 1987

7.10   G. Fielding, M. Elkateb, W. O. Kelham: The application of phase comparison protection to EHV transmission lines; International Conference on Feeder Protection, UMIST, Manchester, 1979

7.11   R. Parmella, A. Stalewski: Voice frequency phase comparison feeder protection; International Conference on Feeder Protection, UMIST, Manchester, 1979

7.12   GEC Measurements Publication: Phase comparison carrier protection Type Contraphase P10

7.13   Siemens Publication: Statischer Phasenvergleichsschutz Type 7SD31

7.14   K. Schneider, G. Ziegler: Modular phase comparison with PLC or fibre optical link, IEE Conf. Publ. No. 249, London 1985, P. 56-59

7.15 R. Requa: Die Grenzen der Anwendbarkeit des Distanzschutzprinzips; ETG-Fachberichte 12 - Selektivschutz, Nurnberg, 1983, P. 22-28

7.16 H. Ungrad, V. Narayan: Behaviour of distance relays under earth fault conditions on double-circuit lines; Brown Boveri Review, No. 10, 1969, P. 494-501

7.17 H. Widmer: Application of a family of modern protection signalling equipment; Brown Boveri Review, No. 8, 1977, P. 446-454

7.18 VDEW-Ringbuch Schutztechnik: Richtlinien für die Schutzsignalübertragung - Übertragungssysteme für Schutzaufgaben, 1986

7.19 H. Ungrad: Distance relays with signal transmission for main and back-up protection; ABB Relays Publication CH-ES 35-30.12 E

7.20 A. Otto, R. Schäfer, S. Reinhard, F. Frey: Detecting high resistance earth faults; ABB Review, No. 1, 1990, P. 19-26

7.21 O. E. Lanz: Eine Auswertung von nichtstationären Signalen für den schnellen Schutz von Hochspannungsleitungen; Dissertation, Technical University of Zurich, 1982

7.22 O. E. Lanz, F. Engler, M. Hänggli. G. Bacchini: LR91 — an ultra high-speed directional comparison relay for the protection of high-voltage transmission lines; Brown Boveri Review, No. 1, 1985, P. 32-36

7.23 R. P. Carter: Ultra high-speed relay for EHV/UHV lines base on directional wave detection principles; IEE Conference Publication No. 185, London, 1980, P. 166-177

7.24 H. Ungrad: Back-up protection; ABB Relays Publication CH-ES 38-01 E

7.25 K. Becker: Der Reserveschutz im Höchstspannungsnetz; ETG-Fachberichte, No. 12, Selektivschutz, Nurnberg, 1983, P. 29-44

7.26 Z. Pilch, W. Winkler: Phase comparison and distance protection performance during simultaneous double faults, IEE Conf. Publ. No. 249, London 1985, P. 42-45

7.27 Richtlinien für die Kurzunterbrechung in elektrischen Netzen; VDEW, Frankfurt, 1981

7.28 F. Ilar: Auto-reclosing; ABB Relays Publication CH-ES 40-20 E

7.29 W. Kolb, I. De Mesmaeker: New devices for auto-reclosure and their applications; Brown Boveri Review, No. 2, 1981, P. 79-86

7.30 R. Nylen: Auto-reclosing; ASEA Journal, No. 6, 1979, P. 127-132

7.31 J. Zydanowicz: Electroenergetique automatique de protection, Volume III: Les automatismes de reprise et de prevention; WNT, Warsaw, 1987

7.32 M. Canay: Stresses in turbogenerator sets due to electrical disturbances; Brown Boveri Review, No. 4, 1975, P. 178-183

7.33 M. Gacka: Schutztechnische Behandlung des Erdschlußfehlers in Mittelspannungsnetzen des Kohlebergbaus; Dissertation University of Dortmund, 1985

7.34 F. Ilar, F. Frey, G. Wacha: Power system protection with 900 series solid-state relays; Brown Boveri Review, No. 10, 1987, P. 579-585

7.35 Siemens Publication: Statisches Erdschlußwischerrelais Typ 7SN7098

7.36 ASEA Relays: Reläskydd för distributionsnät — Applikationsguide AG03 - 7205, 1987

7.37 AEG-Telefunken Publication: Oberschwingungs-Erdschlußrichtungsüberwachungseinrichtung RERO bzw. SERO

7.38 P. Schwetz: Ausgleichsströme beim Erdschluß im gelöschten Netz; Elektrizitätswirtschaft, No. 22, 1980, P. 854-858

7.39 ABB Relays Publication CH-ES 53-10 E: Instructions for planning differential protection schemes

7.40 H. J. Müller: Struktureller Aufbau statischer Schutz- und Sekundäreinrichtungen in Mittelspannungsnetzen; ETG-Fachberichte 12 — Selektivschutz, VDE-Verlag GmbH, Berlin - Offenbach, 1983, P. 60-74

7.41 J. B. Royle, A. Hill: Low impedance biased differential busbar protection for application to busbars of widely differing configuration; IEE Conference Publication No. 302, Edinburgh, 1989, P. 40-44

7.42 G. Böhme, P. Magajna: Plant protection with relays from the Modures system; Brown Boveri Review, No. 8/9, 1981, P. 338-347

7.43 Sammelschienenschutz, VDEW, 1984

7.44 Siemens Publication: Elektronischer Sammelschienenschutz Typ 7SS10

7.45 ASEA Publication: Basic theory of bus differential protection Type RADSS, Pamphlet RK- 637-300E

7.46 AEG-Telefunken Publication: Elektronischer Sammelschienenschutz

7.47 ABB Relays Publication CH-ES 23-62.01 E: Static busbar protection based on the directional comparison principle

7.48 ABB Relays Publication CH-ES 25-64 E: Static busbar protection Type INX5

7.49 W. Winkler: Korrekte Funktionsweise von Schutzeinrichtungen trotz temporärer Stromwandlersättigung; Elektrie, No. 6, 1989, P. 211-213

374   References

7.50   G. Potisk, K. W. Zwahlen: The new breaker failure protection system
        SX91; Brown Boveri Review, No. 10, 1986, P. 575-578

7.51   IEEE Power System Relaying Committee Report: Summary update of
        practices on breaker failure protection; IEEE Transactions on PAS, Vol-
        ume 101, No. 3, 1982, P. 555-563

7.52   ASEA Publication: Breaker failure relaying; Lecture 651-909 E, 1984

Chapter 8

8.1   G. Cotto: Harmonisation of protection policies in power stations and on
        the network; CIGRE, Paris, 1984, Report 34-14

8.2   K. L. Klein: Schutzeinrichtungen großer Kraftwerksblöcke; ETG-Fach-
        berichte 12 - Selektivschutz, Nurnberg, 1983, P. 45-59

8.3   E. Zurowski: Schutz großer Kraftwerksblöcke im Netzbetrieb; etz-Ar-
        chiv, Volume 9, 1976, P. 667-668

8.4   J. Gantner: Protection for turbogenerators; Brown Boveri Review, No. 1,
        1980, P. 22-25

8.5   H. Schaefer: Elektrische Kraftwerkstechnik; Springer Verlag, Berlin -
        Heidelberg - New York, 1979

8.6   Electricity Council: Power System Protection, Volume 3, Application,
        Peter Peregrinus Ltd., Stevenage UK and New York, 1981

8.7   A. M. Dmitrenko, V. M. Kiskatchy, D. D. Levkovich, L. A. Nadel, Y. G.
        Nazarov, V. N. Vavin: Protection of large thermal and nuclear power
        plants with large units; CIGRE, Paris, 1984, Report 34-06

8.8   ASEA Publication: Generator Protection — Application Guide AG03-
        4005E, 1986

8.9   H. Ungrad, V. Narayan, M. Fiorentzis, M. Ilar: Coordinated protection of
        a large power plant and the connected network; CIGRE, Paris, 1978, Re-
        port 34-05

8.10   G. Bär, H. W. Grau, L. Kienast: Der Generatorschutz in elektronischer
        Bauweise im Pumpspeicherwerk Wehr; Elektrizitätswirtschaft, Volume 5,
        1979, P. 147-154

8.11   VDEW-Richtlinien für den elektrischen Blockschutz, 1985

8.12   ABB Relays Publication CH-ES 31-01 E: The selection of generator
        protection system

8.13 A. Fischer: Statischer Generatorschutz in Subsystemtechnik; Siemens Energietechnik, Volume 6, 1984, P. 282-285

8.14 ABB Relays Publication CH-ES 63-20 E: Solid-state generator protection modular system GSX5e

8.15 Siemens Publication: Statischer Maschinenschutz — Gerichteter Erdschlußschutz Typ 7SN33

8.16 AEG-Telefunken Publication: Ständererdschlußschutz für Maschinen im Sammelschienenbetrieb in isolierten bzw. kompensierten Netzen

8.17 L. Fickert: Ständererdschlußerfassung bei Generatoren mit hochohmiger Sternpunktserdung; Elin-Zeitschrift, Volume 1, 1979, P. 15-18

8.18 M. Ilar, J. Zidar, M. Fiorentzis: Innovations in generator protection; Brown Boveri Review, No. 6, 1978, P. 379-387

8.19 J. W. Pope: A comparison of 100 % stator ground fault protection schemes for generator stator windings; IEEE Transactions on PAS, No. 4, 1984, P. 832-840

8.20 L. Pazmandi: Protection contre les défauts à la terre du stator des grands alternateurs; CIGRE, Paris, 1972, Report 34-01

8.21 J. Zydanowicz: Elektroenergetyczna automatyka zabezpieczeniowa, T.2 - Automatyka eliminacyjna; WNT Warsaw, 1985

8.22 E. F. Knütter, K. Nimes: Neuartiger Ständererdschlußschutz für Hochspannungsmaschinen; Siemens Zeitschrift, Volume 12, 1972, P. 909-911

8.23 Siemens Publication: Statischer Maschinenschutz — Ständererdschlußschutz 7UE22

8.24 M. Ilar, V. Narayan, F. von Roeschlaub, C. J. Pencinger: Total generator ground fault protection; Georgia Technical Protective Relaying Conference, Atlanta, 1979

8.25 AEG-Telefunken Publication: Maschinenschutz — Generator-Differentialschutz SQG

8.26 ASEA Publication: High-impedance three-phase differential relay Type RADHA

8.27 ABB Relays Publication CH-ES 65-25 E: Relays for high-impedance differential protection Types UZ91 and UZ92

8.28 T. Forford: High-speed differential protection for large generators; IEE Conference Publication No. 185, London, 1980, P. 30-33

8.29   F. Peneder, H. Butz: Exciter systems for three-phase generators in indus-
       trial and medium-size power stations; Brown Boveri Review, No. 1,
       1974, P. 41-50

8.30   F. Peneder, H. Butz, M. Fiorentzis: Protection of industrial generators
       and industrial networks taking the excitation system into consideration;
       Brown Boveri Review, No. 1, 1974, P. 36-40

8.31   ABB Relays Publication CH-ES 62-81 E: Underimpedance relay Type
       ZSX102

8.32   Siemens Publication: Maschinen-Impedanzschutz Typ 7SL15

8.33   J. Gantner, A Schwengeler, P. P. Warto: Protection for large turbogen-
       erators; Brown Boveri Review, No. 9, 1975, P. 428-434

8.34   Siemens Publication: Statischer Maschinenschutz — Querdifferential
       schutz Typ 7UQ21

8.35   S. Wroblewska, H. Dytry: Criteria for detection of single earth faults in
       generator excitation circuits; Energetyka, No. 1/2, 1974, P. 14-16

8.36   E. F. Knütter, G. Ziegler: Elektronischer Läufererdschlußschutz mit
       neuartigem Meßprinzip; Siemens Zeitschrift, Volume 12, 1972, P. 906-
       909

8.37   K. Henninger, E. F. Knütter, K. Schmiedel: Zweistufiger Läufererd-
       schlußschutz hoher Empfindlichkeit für Synchronmaschinen; Siemens En-
       ergietechnik, Volume 11, 1979, P. 407-411

8.38   AEG-Telefunken Publication: Maschinenschutz — Läufererdschlußschutz
       SLG

8.39   ABB Relays Publication CH-ES 62-15.1 E: Negative-sequence relay
       Type IPX146

8.40   J. A. Imhof: Out-of-step relaying for generators; WG report, IEEE Power
       System Relaying Committee Report, IEEE Transactions on PAS, No. 5,
       1977, P. 1556-1564

8.41   B. Gaillet, J. Gantner: Protection against out-of-step operation of large
       synchronous machines; CIGRE SC34, WG-01 Report, Electra, Volume
       50, 1974, P. 77-92

8.42   A. Stalewski, J. L. H. Goody, J. A. Downes: Pole-slipping protection;
       IEE Conference Publication No. 185, "Developments in power system
       protection", London, 1980, P. 38-44

8.43   H. Ungrad, W. Winkler: Problems and trends in protection of large power
       station units; Energetyka, No. 9, 1978, P. 354-357

8.44 M. S. Baldwin, W. A. Elmore, J. J. Bonk: Improved turbine/generator protection for increased plant reliability; IEEE Transactions on PAS, No. 3, 1980, P. 982-988

8.45 Siemens Publication: Außertrittfallschutz für elektrische Maschinen Typ 7VM30

8.46 W. R. Roemish, E. T. Wall: A new synchronous generator out-of-step relay scheme; IEEE Transactions on PAS, No. 3, 1985, Part I, P. 563-571, Part II, P. 572-582

8.47 F. Ilar: Innovations in the detection of power swings in electrical networks; Brown Boveri Review, No. 2, 1981, P. 87-93

8.48 American Standard requirements for cylindrical rotor synchronous generators; ANSI C.50.13

8.49 Siemens Publication: Statischer Maschinenschutz-Überlastschutz 7SK24 mit 2-kanaligem thermischem Abbild

8.50 V. Narayan, H. Schindler, C. J. Pencinger, D. Carreau: Frequency excursion monitoring of large turbogenerators; IEE Conference Publication No. 185, London, 1980, A45-48

8.51 H. Ungrad: Design of system protection in connection with large power station units; etz-Archiv, Volume 9, 1976, P. 535-537

8.52 H Fick, A. Fischer: Kraftwerkentkupplung, ein Schutzgerät zur Vermeidung hoher Torsionsbeanspruchungen von Turbosätzen; Siemens Energietechnik, Volume 12, 1979, P. 431-435

8.53 J. Berdy, P. H. Brown, C. A. Mathews, D. N. Walker, S. B. Wilkinson: Automatic high-speed reclosing near large generating stations; CIGRE, Paris, 1982, Report 34-02

8.54 A. Fischer: Standardisierter Blockschutz von Gasturbinenkraftwerken; Siemens Energie & Automation, No. 3, 1987, P. 28-31

8.55 A. Fischer: Übererregungsschutz für Transformatoren; Siemens Zeitschrift, No. 1, 1976, P. 20-22

8.56 J. Gantner, F. H. Birch: Transformer overfluxing protection; Electra, No. 31, 1973, P. 65-73

8.57 ABB Relays Publication CH-ES 31-03 E: Modular generator protection GSX

8.58 P. Magajna, M. Franzl, J. Zidar: Motor protection equipment; Brown Boveri Review, No. 6, 1978, P. 404-412

8.59 VDEW-Richtlinien für den Schutz von Hochspannungs-Asynchron-motoren in Kraftwerken, 1984

8.60 ASEA Publication RK 642-300E: Motor protection Type RAMDA

8.61 W. Amey, K. Nimes: Erweiterter elektronischer Überlastschutz 7SK2 für große Drehstrommmotoren; Siemens Zeitschrift, Volume 8, 1976, P. 554-557

8.62 A. N. Eliasen: The protection of of high-inertia drive motors during abnormal starting conditions; IEEE Transactions on PAS, No. 4, 1980, P. 1462-1471

8.63 S. E. Zocholl, E. O. Schweitzer, A. Aliaga-Zegarra: Thermal protection of induction motors enhanced by interactive electrical and thermal models; IEEE Transactions on PAS, No. 7, 1984, P. 1749-1755

Part C (Chapters 9 to 13)

C.1 A. Antoniou: Digital filters analysis and design; McGraw-Hill, 1979

C.2 P. Bornard, J. C. Bastide: A prototype of a multiprocessor based distance relay; IEEE Transactions on PAS, Volume 101, No. 2, 1982, P. 491-498

C.3 T. Cholewicki: Elektrotechnika teoretyczna, Volume 2, WNT Warsaw, 1971

C.4 A. J. Degens: Microprocessor implemented digital filters for calculation of symmetrical components; IEE Proceedings, Pt. C, No. 3, 1982, P. 111-116

C.5 L. Erickson, M. M. Saha, G. D. Rockefeller: An accurate fault locator with compensation for apparent reactance in the resistance resulting from the remote end infeed; IEEE Transactions on PAS, Volume 104, No. 2, 1985, P. 424-436

C.6 C. B. Gilcrest, G. D. Rockefeller, E. A. Udren: High-speed relaying using a digital computer; IEEE Transactions on PAS, Volume 91, No. 3, 1972, P. 1235-1258

C.7 Takeshi Hayashi, Tetsuo Kona: New protection algorithm for a digital control system; CIGRE SC34, Colloquium, Philadelphia, 1975

C.8 IEEE Power System Relaying Committee WG Report: Criterion for the behaviour of the digital impedance method of transmission line protection; IEEE Transactions on PAS, Volume 104, No. 1, 1985, P. 126-135

C.10 B. Jeyasurya, W. J. Smolinski: Identification of the best algorithm for digital distance protection of transmission lines; IEEE Transactions on PAS, Volume 102, No. 10, 1983, P. 3358-3369

C.11 B. Jeyasurya, T. H. Vu, W. J. Smolinski: Determination of transient apparent impedances of faulted transmission lines; IEEE Transactions on PAS, Volume 102, No. 10, 1983, P. 3370-3378

C.12 A. T. Johns, M. A. Martin: New ultra high-speed distance protection using finite transform technique; IEE Proceedings, Pt. C, No. 3, 1983, P. 127-138

C.13 B. P. Lathi: Systemy telekomunikacyjne; WNT Warsaw, 1972

C.14 B. J. Mann, I. F. Morrison: Digital calculation of impedance for transmission line protection; IEEE Transactions on PAS, Volume 90, No. 2, 1971, P. 270-279

C.15 M. A. Martin, A. T. Johns: New approach to digital distance protection using time graded impedance characteristics; Proc. of 16 UPEC, Sheffield, 1981, Report A.5.3

C.16 P. C. Mc Laren, M. A. Redfern: Fourier series technique applied to distance protection; IEEE Transactions on PAS, Volume 94, No. 11, 1975, P. 1301-1305

C.17 A. V. Oppenheim, R. W. Schafer: Cyfrowe przetwarzanie sygnałów; WKL Warsaw, 1978

C.18 A. G. Phadke, M. Ibrahim, T. Hlibka: Fundamental basis for distance relaying with symmetrical components; IEEE Transactions on PAS, Volume 96, No. 3, 1977, P. 635-646

C.19 A. G. Phadke, T. Hlibka, M. Ibrahim, M. G. Adamiak: A microcomputer based symmetrical component distance relay; IEEE PICA Conference, 1979, 79CH 1301-3PWR, P. 47-55

C.20 A. G. Phadke, J. S. Thorp, M. G. Adamiak: A new measuring technique for tracking voltage phasors, local system frequency and rate-of-change frequency; IEEE Transactions on PAS, Volume 102, No. 5, 1983, P. 1025-1033

C.21 Final report on computer based protection and digital techniques in substations: CIGRE CE/SC 34 GT/WG 02, 1985

C.22 R. Poncelot: The use of digital computers for network protection; CIGRE, Paris, 1970, Report 32-08

C.23 A. M. Ranjbar, B. J. Cory: An improved method for the protection of high voltage transmission lines; IEEE Transactions on PAS, Volume 94, No. 6, 1974, P. 1522-1534

C.24 M. S. Sachdev, M. A. Baribeau: A new algorithm for digital impedance relays; IEEE Transactions on PAS, Volume 100, No. 12, 1981, P. 4815-4820

C.25 M. S. Sachdev, M. M. Giray: A least square technique for determining power system frequency; IEEE Transactions on PAS, Volume 104, No. 1, 1985, P. 437-444

C.26 T. Takagi, Y. Yamakoshi, M. Yamaura, R. Konow, T. Matsuhima: Development of a new type fault locator using the one-terminal voltage and current data; IEEE Transactions on PAS, Volume 101, No. 8, 1982, P. 2892-2898

C.27 J. S. Thorp, A. G. Phadke: A microprocessor based three-phase transformer differential relay; IEEE Transactions on PAS, Volume 101, No. 2, 1982, P. 426-432

C.28 A. Wiszniewski: Accurate fault impedance locating algorithm; IEE Proc. Pt. C, No. 6, 1983, P. 311-314

C.29 A. Wiszniewski: New algorithm for calculating current and voltage for fast protection; IEE Proc. Pt. C, No. 1, 1987, P. 81-82

C.30 A. Wiszniewski: Digital algorithms for differential protection of power transformers; Proceedings of the 9th. PSCC, Cascais, Protugal, 1987, P. 725-731

C.31 Q. S. Yang, I. F. Morrison: Microcomputer based algorithm for high resistance earth fault distance protection; IEE Proc. Pt. C, No. 6, 1983, P. 306-310

C.32 Digital protection techniques and substation functions: CIGRE Publications, CE/SC 34 GT/WG 01, 1989

C.33 Y. V. V. S. Murty, W. J. Smolinsky: Design and implementation of a digital differential relay for a three-phase power transformer based on the Kalman filtering theory; IEEE Transactions on Power Delivery, Volume 3, No. 2, 1988, P. 525-533

C.34 S. H. Horowitz, A. G. Phadke, J. S. Thorp: Adaptive transmission system relaying; IEEE Transactions on Power Delivery, Volume 3, No. 4, 1988, P. 1436-1445

C.35 C. Christopoulos, D. W. P. Thomas, A. Wright: Signal processing and discriminating techniques incorporated in a protective scheme based on travelling waves; IEE Proc. Pt. C, No. 4, 1989, P. 279-288

C.36 R. K. Aggarwal, A. T. Johns: A differential line protection scheme for power systems based on composite voltage and current measurements; IEEE Transactions on Power Delivery, Volume 4, No. 3, 1989, P. 1595-1601

C.37 G. Benmouyal: Design of a universal protection relay for synchronous generators; CIGRE, 1988, Paris, Report 34-04

C.38 A. Engquist, L. Eriksson: Numerical distance protection for sub-transmission lines; CIGRE, 1988, Paris, Report 34-04

C.39 O. E. Lanz, W. Fromm: A new approach to digital protection; CIGRE, 1988, Paris, Report 34-12

C.40 I. Korbasiewicz, M. Korbasiewicz, W. Winkler: A microprocessor based protective system for generator/transformer units; IEE Conference Publication No. 302, Edinburgh, 1989, P. 56-60

C.41 A. Wiszniewski, J. Szafran: Distance digital algorithm immune to saturation of current transformers; IEE Conference Publication No. 302, Edinburgh, 1989, P. 196-199

C.42 D. M. Peck, F. Engler, I. De Mesmaeker: A second generation micro-processor line protection relay; IEE Conference Publication No. 302, Edinburgh, 1989, P. 200-204

C.43 G. Koch, G. Ziegler: Modern practices and field experience with MC-based relays; IEE Conference Publication No. 302, Edinburgh, 1989, P. 315-320

C.44 An international survey of the present status and the perspective of expert systems in power system analysis and techniques: CIGRE Publications, CE/SC 38 GT/WG 02 TF 07, 1988

C.45 J. Trecat, Jianping Wang: Expert system application in substation monitoring, back-up protection and control; IEE Conference Publication No. 302, Edinburgh, 1989, P. 75-79

C.46 L. L. Lai: Development of an expert system for power system protection coordination; IEE Conference Publication No. 302, Edinburgh, 1989, P. 310-315

C.47 A. Wiszniewski: Digital high-speed calculation of the distorted signal fundamental component; IEE Proc. Pt. C, No. 1, 1990, P. 19-24

Index

382